INCREDIBLE WOMEN OF INDIA'S FREEDOM STRUGGLE

INCREDIBLE WOMEN OF INDIA'S FREEDOM STRUGGLE

Suneeta Koundal

PENTAGON PRESS LLP

First published in 2023 by

PENTAGON PRESS LLP
206, Peacock Lane, Shahpur Jat
New Delhi-110049, India
Contact: 011-64706243

Typeset in Palatino, 11 Point
Printed by Aegean Offset Printers, Greater Noida, U.P.

ISBN: 978-93-90095-65-0 (HB)

www.pentagonpress.in

To all those who sacrificed their lives for the freedom of our country—men, women, boys and girls. You are and will always be cherished.

CONTENTS

Foreword *ix*

Preface *xi*

Acknowledgements *xv*

1. Velu Nachiar 1
2. Rani Kittur Chennamma 6
3. Begum Hazrat Mahal 11
4. Rani Lakshmi Bai 18
5. Kuyili 24
6. Uda Devi 28
7. Jhalkari Bai 32
8. Annie Wood Besant 37
9. Bhikaji Rustom Cama 44
10. Kasturba Gandhi 52
11. Matangini Hazra 57
12. Bibi Gulab Kaur 62
13. Sarojini Naidu 66
14. Basanti Devi 72
15. Umabai Kundapur 76

16.	Ramadevi Choudhury	82
17.	Kamala Nehru	88
18.	Maniben Patel	93
19.	Kamaladevi Chattopadhyay	97
20.	Vijaya Lakshmi Pandit	106
21.	Usha Mehta	111
22.	Satyavati Devi	118
23.	Sucheta Kriplani	122
24.	Tara Rani Srivastava	128
25.	Prabhavati Devi	132
26.	Pritilata Waddedar	135
27.	Aruna Asaf Ali	141
28.	Durgawati Devi	148
29.	Bina Das	155
30.	Freda Bedi	161
31.	Lakshmi Sahgal	169
32.	Rani Gaidinliu	175
33-34.	Santi Ghose and Suniti Choudhury	181
35.	Kanaklata Barua	191
36.	Parbati Giri	198
	Website References	204
	Index	209

FOREWORD

I know writing is difficult even for professional authors who do it all the time and when I was requested to write a foreword for the work done by Suneeta Koundal, which covers the contributions made by these incredible women to India's Freedom Struggle, I found myself incapable and inadequate to pay a fitting tribute to these extraordinary women whose supreme sacrifices cannot be described in words. It is a feeling that shakes our inner conscience and takes us to feel the white light of spirituality and meditation which might have been an inbuilt and inherent instinct of these fiery women who gave their lives and comfort for the freedom of the Motherland.

The adventures recorded in this book are thrilling and one or two are my own experiences, creating a special bond with female freedom fighters. I thank the author for bringing these stories of valour and bravery of these women into the limelight which is greatly needed today when the sacrifice of such heroes is lost in the dim pages of history. It will make the present generations aware about the contribution of these women to the struggle for India's Independence.

Women in India have always been inspiring, in the past during the freedom struggle and also in modern times. However, little is known about them whereas the untold sacrifices made by them should be occupying the foremost place in the country's roll of honour. The long story of struggle and hardship of the Indian freedom movement is incomplete without mentioning these women fighters who fought with a true spirit of patriotism, lived in extremely difficult situations, abandoned their comfort and chose a difficult stormy path, even unto death to enable India to stand with head held high today in the comity of nations.

My salute to these unsung female heroes. My special thanks to the author for compiling these captivating stories into the book that is now in your hands. I like how she has presented the stories in a manner that allows the reader to extract the strong messages therein that will help to choose the life we want for ourselves and live it in the best manner possible. I have delivered numerous lectures on 'Leadership', 'Motivation' and 'Risk Management' in various government, private, and educational institutions in India and abroad and promoting education and adventure sports, I am confident this book will be an asset in motivating youth in the national interest. This book will surely prove to be a source of motivation, encouragement, and enthusiasm for the present and forthcoming generations on whose shoulders the future of Indian society and that of India rests.

Padam Shri Santosh Yadav
Mountaineer

PREFACE

A strong woman can be a stoic warrior and could also become an emotional being who cries at a loss, or a homely woman sitting within the four walls of her home, can throw heavy punches when experiencing rage. Some women are remembered for a short period, some are forgotten. Some re-emerge. Those who re-emerge wrestles with the forces of adversity. No one tells their story or writes about them. While remembering such women who break new ground and leave their footprints on the sands of time, we can never forget those incredible women who are not only a source of inspiration but are idols of womanhood. When one glances into the history of the Independence Movement of India, the fact that gains prominence is that thousands of women also played a revolutionary role in the freedom struggle. Mahatma Gandhi's call awakened so many women, who jumped into the battlefield with tremendous courage and vigorously sacrificed their comforts and luxuries. Their work and sacrifice provide a rich insight into our glorious past.

When I set out to write this book, I realised that women had played a major part in the freedom struggle and were ardent revolutionaries, taking up guns, trained themselves in the art of fighting, formed secret organisations, led protests, revolted against social norms, produced anti-British literature, faced brutal lathi charges, were tormented, shot dead and imprisoned for years. With sheer physical and mental strength, these strong female fighters also fought against patriarchy. Of course, we have heard of some of the famous women freedom fighters, but it was not until I began to research that I discovered that women not just from royal and noble families, but also from vulnerable groups, played crucial roles in the freedom struggle. All of them were willing to sacrifice their lives for a better life, not only for themselves or for those around them but for generations to come. Do we pause to consider the sacrifices they made, and what it was like to be treated as second-class citizens in our own country. It is time that they too receive their fair share of fame, honour, and recognition. **No nation can progress without the effective participation and contribution of women. When we recall our history of nationalism, we cannot ignore the leading roles played by the women pioneers of our freedom struggle.**

This book is a tribute to the women who fought and died for us to enjoy the freedom that we have today. From the multitude of examples, I have attempted to focus on the majority of women freedom fighters from mainstream and humble backgrounds, who were filled with patriotic fervour and eager to give up their lives for freedom.

Most of us are aware of the contributions of prominent women such as Rani Lakshmi Bai, Aruna Asaf Ali, Vijaya Lakshmi Pandit, Sarojini Naidu, Sucheta Kripalani, and Bhikaji Rustom Cama and others. However, only a few people are aware of Kanaklata Barua's sacrifice as an adolescent martyr. Every narrative is enthralling. Many revolutionaries like Pritilata Waddedar and Matangini Hazra were wounded in battle and decided to die rather than be captured by the British. Uda Devi, a Dalit lady disguised as a male, avenged her husband's murder. In 1857, the call for rebellion against the British was growing, and Devi displayed her wrath at Sikander Bagh. The stories of Uda Devi, Kuyili, Begum Hazrat Mahal, Rani Kittur Chennamma, Queen Velu Nachiar, Rani Gaidinliu and others have been mostly overlooked. Jhalkari Bai was a Dalit woman warrior who rose to become one of Rani Lakshmi Bai's most trusted aide. Satyavati Devi, Kamala Devi Chattopadhyay, Parbati Giri, Tara Rani Srivastava, Maniben Patel and others showed an extraordinary level of composure and bravery. Women like Annie Besant and Freda Bedi crossed the borders and contributed vehemently to the Indian independence. Stories of Bina Das, Durgawati Devi, Basanti Devi, Gulab Kaur, Prabhavati Devi, Ramadevi Choudhury, and Uma Bai Kundapur showed the undaunted, unbroken spirit of these remarkable women. Kasturba Gandhi, Kamala Nehru, Lakshmi Sahgal, and Usha Mehta are other names who listened to their instincts and diminished their doubts and fears. Fourteen-year-old Santi Ghose and Suniti Choudhury killed Stevens and were awarded life imprisonment. Those women proved their mettle as they overpowered their emotions with a lot of adjustments.

They accompanied Mahatma Gandhi on the final leg to Dandi, where he picked up a handful of salt in protest against the Salt Tax. In the months that followed, women converged on the streets in support of the freedom struggle. Women from all three presidencies marched to the nearest sea coast to collect salty clay. They organised Prabhat pheris, or morning processions, in Mumbai and Ahmedabad, where they sang patriotic songs. Think about Begum Hazrat Mahal of Awadh who fought hard to protect Lucknow before escaping to Nepal, where she died in 1879. Earlier in 1829, the courageous Rani Kittur Chennamma died in British captivity. These female warriors were essentially fighting British attempts to annex their territories. The women revolutionaries of the early twentieth century, on the other hand, had a greater goal: they wanted to free India from British control and were willing to pay the ultimate price.

There may be many more legendary women I may have overlooked due to lack of knowledge and information, but their sacrifice was not in vain, and they also deserve high recognition. I salute all of them. There is no deliberate intention to promote or ignore any particular person, group, community, caste, creed, culture, religion, etc. This is solely an idea that came straight from my heart and soul.

I have compiled my list of woman freedom fighters and researched each one from a variety of sources, including various libraries, websites, online articles, newspapers and encyclopedias, wikipedia, biographies, autobiographies, and the books & literature available. I am genuinely grateful to each one of these resources and the persons behind them. I have attempted to include all their experiences, crucial dates and major events, but hold myself entirely to blame for any errors or omissions, if any, and will definitely correct it, if brought to my notice. My purpose is to share some compelling stories about these brave, empowered and selfless women, without whom our country would not be what it is today.

SUNEETA KOUNDAL

ACKNOWLEDGEMENTS

It was more difficult than I expected to get an idea and develop it into a book. Whatever we set out to do is done with the support and encouragement of people willing to guide us and share their knowledge. This topic has been close to my heart, and it took me a lot of work and encouragement to see it through. A number of incredible, thoughtful, and loving people were with me at various stages.

I want to thank Almighty God, who is always with me and helps me through my difficulties.

I want to thank my father, late Sh. Rikhi Ram Koundal, a Retd Officer from Ministry of Defence. He has been a source of motivation throughout my life, and continues to do so despite that he is no longer among us.

I owe special thanks to my mother, Smt. Nirmala Koundal, now settled at Kangra, Himachal Pradesh who raised me, loved me and without whose trust, I would be unable to achieve whatsoever I set out to do.

I am deeply grateful to Sh. Sanjay Koundal, my elder brother, settled at Kangra, Himachal Pradesh who encouraged me, provided resources, and stood by me when things looked grim.

My younger brother, Dr. Pawan Koundal, Associate Professor, IIMC deserves special recognition. Where would I be without his unwavering support, academic ideas, and motivation at every step of my life.

My special gratitude to Dr. Vineet Joshi IAS former Chairman CBSE (presently, Addl. Secretary, Ministry of Education, Govt. of India), a well known educational administrator who gave me wings to fly by giving me challenging assignments in my formative years in CBSE, introducing me to the wonderful world of examinations and academics.

My gratitude to my mentor Sh. P.I. Sabu (Retd Director CBSE), a good friend and an officer with great ideas, who always encouraged his colleagues to write books on their works or on any matter as he believes that from these materials future generations will learn and evolve. I acknowledge his encouragement and support throughout the process even when this book was germinating in my head.

My special thanks to my sisters, brothers-in-law, sisters-in-law, niece, and nephews, whose beautiful smiles, adorable chuckles and endless energy brightened me.

I want to thank everyone who helped me create this book—my editor, my designers and specially Shri Rajan Arya, CEO of Pentagon Press LLP who advised me with important tips in my journey as a writer.

I am grateful to Padam Shri (Smt.) Santosh Yadav, a mountaineer having world records in her credit, Prof. Smita Mishra and Prof. Ujjwal Kumar Singh both renowned Academicians for their guidance, reviews and recommendations. They have always been available to help me with my research whenever I needed them. Working with them was a great learning. They not only guided me but also gave me a vision to look beyond the horizons.

Last, but not the least, my dear readers deserve my heartfelt appreciation for choosing to read this book and inviting me into their lives. Thank you!

VELU NACHIAR
(1730–1796)

1

VELU NACHIAR

Woman who wrestled with them "forces"

Velu Nachiyar was born 293 years ago and was India's first female liberation warrior. This lesser-known Tamil Nadu queen fought the British in 1780, 77 years before the First War of Independence. Popularly referred to as 'Veeramangai' (brave woman), with her Dalit commander-in-chief, Velu Nachiyar devised the first suicide bombing in history. She underwent swordsmanship and mace instruction. The princess was trained in a variety of martial arts styles, including Valari and Silambam, and was an exceptional horsewoman and archer. She mastered English, French and Urdu in those times.

Velu Nachiyar was born in a royal family in 1730 and was the only child of Raja Chellamuthu Vijayaragunatha Sethupathy and Rani Sakandimuthal. The royal couple reared her similar to how they would raise a male heir. She received training in horseback riding, archery and martial arts skills like Valeri (throwing sickle) and silambam (fighting with a stick). She possessed fluency in several languages, including English, French and Urdu. When Velu Nachiyar was sixteen years old, in 1746, she married Muthuvaduganathur Udaiyathevar, the prince of Sivagangai. They had a daughter named Vellachi.

Velu was drawn into battle with the British when her husband was killed. The British attacked Sivagangai with the son of the Nawab of Arcot and killed Velu Nachiyar's husband in the 'Kalaiyar Koil War'. She was forced to escape her kingdom with her daughter and found refuge at Virupachi, Dindigul, where she remained for eight years under the sanctuary of Gopala Nayaker. Later, she was joined by Rani Velu Nachiyar's bodyguards Vellaimaruthu and Chinnamaruthu. Rani Velu Nachiyar, in collaboration with the Maruthu brothers, established a new regiment capable of driving the British out.

In 1780, Velu Nachiyar bravely defeated the British in a well-formulated coup. Her intelligence-gathering agents identified the location where the British stored their ammunition. Then it was planned to eliminate all of it with a suicide attack. The commander-in-chief, Kuyili, volunteered to carry out the suicide mission. She soaked herself in ghee and set herself on fire before entering the ammo shop and destroying everything. Many believe Kuyili is Velu Nachiyar's adopted daughter and the world's first documented suicide bomber. Udaiyaal was also Velu Nachiyar's adopted daughter. She, too, died in one of the battles against the British while blowing up their armory in another suicide attack. The queen established an all-women army in her honour and named it Udaiyaal.

Following her victory over the British, Velu Nachiyar ruled the Sivangangai kingdom for ten years. Her daughter Vellachi was made heir to the throne. She delegated administrative powers to the Marudu brothers in 1780. Velu Nachiyar built a mosque and church in Saragani to show her appreciation to Haider Ali for his invaluable and timely support. Vellachi, Velu Nachiyar's daughter, succeeded her on the throne in 1790 and ruled until 1793. The 66-year-old heroic queen passed away on 25 December 1796 in Sivagangai.

Her story was one of heroism and it awarded India its first female fighter. She deserved to be commended for her bravery and for breaking down gender barriers as early as the 18th century. On December 31, 2008, a commemorative stamp was issued in her honour. A Tamil-American hip-hop artist Professor A.L.I. dedicated a song titled 'Our Queen' to Velu Nachiyar as part of his album 'Tamilmatic'. The OVM Dance Academy of Chennai staged a magnificent ballet performance. It recounted the lionhearted queen's life. Another spectacular ballet performance was presented at Chennai's Naradha Gana Sabha. Sriram Sharma, the performance director, conducted an extensive study on her. On 18th July 2014, the late J. Jayalalitha, former Chief Minister of Tamil Nadu, dedicated the Veeramangai Velu Nachiyar

Memorial at Sivagangai. A six-foot bronze statue of the queen was also unveiled, and it was stated that January 3 would be observed as the intrepid queen's birth anniversary.

One must know their strengths and potential. That is how Velu Nachiar used her unique upbringing that molded her into a multifaceted personality and a warriorhood. Her relentless quest, earnestness to retake the town, braced for her troops to scale the fort wall and attack. She is a woman who made other women a precious jewel of her crown.

RANI KITTUR CHENNAMMA
(1778–1829)

2

RANI KITTUR CHENNAMMA

The Queen who was not afraid of fate and fight

Rani Kittur Chennamma was one of the first Indian monarchs to launch an armed revolt against the British East India Company in 1824, protesting the enforcement of the Doctrine of Lapse. Although her rebellion against the British resulted in arrest, she became a legendary freedom fighter in the state of Karnataka and an icon of India's independence. Rani Chennamma epitomized empowerment, courage, and feminism by taking on the might of the British Empire alone, during a time when women were confined indoors.

Kittur Chennamma was born on October 23, 1778, in Kakati, a tiny village in Karnataka's current Belagavi District (56 years before the 1857 revolt led by Rani Lakshmi Bai). She belonged to the Lingayat community. She married Mallasarja Desai, the Kittur king. Her married life was a sad tale as her husband died in 1816. Fate played a tragic game, her only son from her marriage also died in 1824. As a queen of Kittur, Chennamma adopted a boy named Shivalingappa and proclaimed him heir to continue the family tradition.

The British East India Company did not take Chennamma's actions lightly and ordered Shivalingappa's exile from the kingdom. This was done under the pretext of the Doctrine of Lapse, which said that adoptive children of local kings were not permitted to be called their successors and that if the native rulers did not have children of their own, their kingdom would be annexed to the British Empire as a province. The Doctrine of Lapse was codified by Lord Dalhousie between 1848 and 1856.

Rani Chennamma, on the other hand, refused to comply with the British command to depose Shivalingappa from the throne. Rani Chennamma wrote to the Governor of Bombay (Mumbai) to request him to consider the case of Kittur, but Lord Elphinstone refused her request. St. John Thackeray was in charge of the Dharwad collectorate and Mr Chaplin was the commissioner, and the state of Kittur came under the control of the Dharwad collectorate. Both men refused to recognise Rani Chennamma as the regent or Shivalingappa as the sovereign, and they informed Rani Chennamma that she must renounce her kingdom. However, she rejected the British order and continued to rule. It was this that triggered the outbreak of a conflict.

Despite numerous attempts, the British were unable to plunder Kittur's wealth and diamonds, which were estimated to be worth approximately 15 lakh rupees. They had launched an attack on Kittur with a force of 20,000 troops and 400 guns, the majority of whom were drawn from the third troop of the Madras (Chennai) Native Horse Artillery.

The British suffered terrible losses in the first combat between the British and the Kittur, which took place in

October of 1824. During this first encounter, the Kittur soldiers also killed St. John Thackeray, a British collector and political agent who was working for the British government. Rani Chennamma's lieutenant, Amatur Balappa, was mostly responsible for Thackeray's death as well as the losses suffered by the British forces throughout the campaign. Rani Chennamma's forces also captured and imprisoned two British officers, Sir Walter Elliot and Mr Stevenson. Rani Chennamma talked with the British Commissioner Mr Chaplin and the Governor of Bombay, under whose government Kittur fell, to avert more destruction and conflict. She freed the hostages in exchange for a British assurance that the war would be over. However, the pledge proved to be a ruse. Mr Chaplin, humiliated by their initial loss at the hands of a minor Indian king, returned with significantly greater armies from Mysore and Sholapur to attack Kittur once more.

With the assistance of her lieutenant Sangoli Rayanna and Gurusiddappa, Rani Chennamma fought the second fight valiantly. Mr Munro, Sir Thomas Munro's nephew, was also killed during this second phase of combat. Rani Chennamma and her warriors held their fort tenaciously for twelve days, but Rani Chennamma fell victim to deception once again. Mallappa Shetty and Vankata Rao, two soldiers in Rani Chennamma's army, betrayed her by mixing mud and cow dung with the gunpowder used in the canons.

Rani Chennamma and her men were ultimately outnumbered by the British forces. Rani Chennamma was beaten in her final battle and captured by the British, who sentenced her to life imprisonment at the Bailhongal Fort. Sangoli Rayanna, her devoted lieutenant, conducted the guerrilla battle in her absence until 1829, but in vain. He desired to place Shivalingappa, Rani Chennamma's adoptive son, as Kittur's ruler but was apprehended and hanged by the British. British forces also apprehended Shivalingappa.

Rani Chennamma spent the final five years of her life in captivity at Bailhongal Fort, where she read religious texts and performed prayers. On February 21, 1829, she expired at the Bailhongal Fort. Throughout the freedom struggle, her valiant resistance to British soldiers inspired countless inspirational plays, folk ballads (Lavani), and short stories. Rani Chennamma's first triumph over British forces is still commemorated annually in October during Kittur's 'Kittur Utsava'.

B.R. Panthulu produced and directed a historical drama film titled Kitturu Chennamma about the lives and times of Kittur Rani Chennamma. Rani Chennamma Express, a popular daily Indian Railways train connecting Bangalore and Kolhapur, was also named after her. On September 11, 2007, Smt. Pratibha Patil, India's first female President, inaugurated Rani Chennamma's statue at the Indian Parliament complex in New Delhi. The Kittur Rani Chennamma Memorial Committee donated the statue, which was carved by Vijay Gaur. Two additional Rani Chennamma sculptures have been placed in Bangalore and Kittur.

A woman has vigor to bear a child, she can swing a sword too, is what required for fighting such a battle. Warrior queen who led an armed rebellion against the British east India company and gave a lesson that Indian rulers will not accept the enforced laws of British without a good fight.

BEGUM HAZRAT MAHAL
(1820–1879)

3

BEGUM HAZRAT MAHAL

"The Revolutionary Queen of Awadh"

Begum Hazrat Mahal was a politically astute woman. Those who are familiar with Hazratganj in Lucknow will never forget this courageous royal Awadh who came from modest beginnings to become a courtesan, queen, and eventually leader. In Awadh, she led an army revolt of 'peasants in uniform.' It quickly escalated into a popular uprising. She ensured social equality for all castes and religions. Her aggressive operations against the English were directed not only at securing Awadh's independence, but also at securing India's independence. She received accolades for her military and administrative prowess. She commanded the famed Siege of Lucknow Residency firsthand. This Queen fought against the British East India Company's oppressors and, subsequently, Queen Victoria herself. Begum eventually sought asylum in Nepal, where she stayed for more than two decades and died on April 7, 1879.

Begum Hazrat Mahal was born as Muhammadi Khanum in 1820 in Faizabad, Awadh, India, in a poor Syed family. Her father, Umber, is said to have been a slave. She was pushed into the royal Awadh harem as a Khawasin (attendant) at a very young age. There she was taught the royal ways, etiquettes and manners. She was beautiful, intellectual, and extremely inventive, and she quickly made her way up in the ranks of the royal Pari Khana (House of Fairies). Soon she caught the eye of the last Tajdar-e-Awadh, Nawab Wajid Ali Shah. The Nawab was an ardent promoter of the arts and intended to transform Awadh into a cultural and literary hub. He established the House to train young and beautiful girls to become professional singers and dancers.

During this era, even courtesans and dancers were refined and revered. The young Muhammadi was rechristened 'Mahak Pari' and became one of the Nawab's 'Mutah' wives, or 'temporary wife under contract.' It was a legal union based on the Shia interpretation of Islamic marriage. The Nawab had many Mutah wives, and whichever of them gave him a son was recognised as "Begum", and eventually designated as the "official" wife. Muhammadi was bestowed the title of "Iftakkar-un-Nissa," or "Pride of All Women". According to several sources back then, Begum Hazrat Mahal was not much liked in the royal household because she had no royal family connections. Moreover, she was adored by the Nawab for her beauty and talent, which made the other ladies envious of her.

Awadh was a rich source of cotton and indigo as a territory. The East India Company seized the kingdom of Awadh in 1856 and ordered the Nawab to step down from the throne on account of lawlessness and severe revenue mismanagement. Nawab Wajid Ali Shah, the last Nawab of Awadh, was exiled to Calcutta (now Kolkata) with a part of his family. The Nawab had thought that his relocation to Calcutta would be temporary. Furthermore, he also intended to make an official complaint or plea to Queen Victoria to regain his glory. Having a streak of rebellion, Hazrat Mahal wanted her husband to resist and fight on the battlefield for the kingdom. However, her husband, the King of Awadh, gave up the kingdom.

Hazrat Mahal, her son Birjis Qadir and Wajid Ali Shah's other wives and 'Mutah' spouses remained in Lucknow. The controversial "Doctrine of Lapse," was used by the British as a vague reason to annexe kingdoms across the country, and this was also provoking animosity among Indian royals all across. On April 8, 1857, the execution of Mangal Pandey in Barrackpore functioned as a catalyst, and what began as a rebellion swiftly spread like wildfire. This was India's first war of independence, fought in 1857.

The people of Awadh were adamant to rise and fight. They were already furious that their ruler had been ousted and couldn't stand to silently watch rebellious soldiers from Meerut and Bengal (as most British army recruits were from Awadh) pouring in. Attacks on British camps were rising, but there was also a lack of unity and confusion prevailed. At that point, a central authority was required to rally them, as the local Jagirdars and Policemen had joined the rebellion but lacked direction. Since Wajid Ali was not there, the rebels turned to young Brijis Qadr, son of Begum Hazrat Mahal, who was barely 12 years old at the time. Despite being the official Crown Prince, he was too young to lead and it was Begum Hazrat Mahal who took over the administration and strategic planning during this crucial period. She took the lead from the front, shocking everyone because it was customary at the time for women to remain largely hidden behind the Purdah or veil.

Begum Hazrat Mahal recruited troops and joined the cavalry, artillery, and infantry as the King's mother. She urged people from all walks of life to contribute money. She was effective in establishing social equality among people of all classes and religions by constructing a harmonious equation. She formed a high-level counsel for the state's good governance, and members included Mummu Khan, Maharaja Balakrishna, Babu Purna Chand, Munshi Ghulam Hazrat, and Mohammad Ibrahim Khan, Raja Man Singh, Raja Desibaksh Singh, Raja Beni Prasad, and others. Her army was led by King Raja Jai Lal Singh. For roughly 10 months, Hazrat Mahal controlled the state on behalf of her son, challenging the British force by instilling patriotism in the people and fellow local rulers.

Hazrat Mahal inspired the people of Awadh and led a military rebellion of "peasants in uniform" in Awadh. It erupted into a general populist rebellion. She wished for India's independence. She used to ride an elephant and physically be present on the battlefield to urge soldiers fighting against the advancing British army. Nana Sahib (Rani Lakshmi Bai's childhood pal) was one of her confidants. She collaborated closely with him.

Soldiers from various regions of north India took up arms and lay siege to Lucknow's British Residency, where British residents had been hiding for four and a half months between July and November 1857. For a long period, the Begum's warriors kept them in the residency. The British troops who entered the city were unable to achieve anything because of the stiff resistance from Begum's troops. The siege lasted a long time, and she led her forces heroically.

Campbell advanced with 4,500 men on 9 November 1857 to rescue the trapped force at Lucknow. The Dilkusha Palace was stormed on 14 November, and two days later, the rescue force pushed through rebel lines. On 22 November, the men, women, and children trapped at the Residency were evacuated to Cawnpore (Kanpur). Begum battled fiercely against the enemy, but her position weakened. Despite the Begum's opposition, the English Commander was able to conduct the imprisoned garrison out of the residency to Alum Bagh, leading to the deaths and injuries of several British officers.

Begum Hazrat Mahal lost control of Lucknow and the majority of Awadh on 16 March 1858, when British soldiers returned; she fled the state with her army. Her inability to preserve the kingdom, on the other hand, did not discourage her from organising armies in other locations. She soon grew tired of just strategizing and joined the war alongside bravehearts like Rani Lakshmi Bai, Bakht Khan, and Maulvi Abdullah. She also collaborated closely with Nana Saheb and later assisted the Maulavi of Faizabad during the attack on Shahjahanpur.

But, by March 19, 1858, Moosabagh, Char Bagh, and Kesar Bagh were captured by the British. Begum Hazrat Mahal had lost the majority of her supporters by the end of 1859, following a brief homestay in Terai. She, her followers, her son Birjis Qadir and Nana Saheb fled to Nepal to seek sanctuary. The British authorities promised her a large sum of money and nice accommodations in exchange for her return to Lucknow. However, the Begum dismissed them and made it clear that she would accept nothing less than an independent Awadh state. She spent her whole fortune in the country, seeking to provide for the thousands of people who had followed her out of loyalty. In the first war of independence, the indomitable Begum was one of the few women to fight against the British.

She was laid to rest on the grounds of Kathmandu's Jama Masjid. To acknowledge her crucial role in India's first liberation movement, she was officially honoured on 15 August 1962 at the Old Victoria Park in Hazratganj, Lucknow, with a marble plaque displaying the Awadh family's coat of arms. Her nation lived within her as she continued to keep abreast of domestic events, wrote poetry and became active in good causes. Begum Hazrat Mahal established herself as a far superior ruler and leader to the Nawab himself. She demanded dignity, both from her court fans and from her adversaries. She was an insightful politician endowed with extraordinary military and administrative prowess.

Passions alone do not change the fates of nations. For that, one needs a revolution, and that was precisely what was brewing across India at the time. Begum Hazrat Mahal is one who will always be remembered as the bravest and strongest woman warrior of the royal family.

RANI LAKSHMI BAI
(1828–1858)

4

RANI LAKSHMI BAI

When the king is a Woman

Who hasn't heard about Jhansi's courageous Rani Lakshmi Bai and her gallant actions against the British in India? Her real name was Manikarnika, from the Manikarnika Ghat, one of the oldest cremation ghat in Varanasi. Following her marriage, she was given the name Lakshmi Bai. Due to her father's position in court, Rani Lakshmi Bai possessed greater freedom than most women of those times. She mastered self-defense, riding and archery, and even formed her army from female court companions. The gallant queen went on to be one of the most famous figures of the Indian battle for independence, having chosen sacrifice over submission to the British.

Manikarnika, also known as Manu, was born on November 19, 1835 in Benares (now Varanasi). Her mother died when she was four years old, and her father, Morepant Tambe, relocated to Bithoor, near Kanpur, to live with his cousin, a former Peshwa of the Maratha Empire. When Manu was thirteen years old, her father, Moropant Tambe, worked at the court of Peshwa Baji Rao II before travelling to the court of Raja Gangadhar Rao Newalkar, the Maharaja of Jhansi. Nana Sahib, Peshwa Baji Rao II's adopted son, and Tantia Tope were two small boys, who were Manu's childhood playmates. Manu was given battle training alongside Nana and Tope as a result of her upbringing, and she excelled at both. She mastered self-defence, riding, and archery and even raised her army of female courtiers. She married Gangadhar Rao, Raja of Jhansi. Raja Gangadhar Rao was incredibly generous and sympathetic. He was an excellent administrator in Jhansi. During his reign, the people of Jhansi were immensely happy.

Rani Lakshmi Bai gave birth to a boy in 1851 who died when he was only four months old. They adopted Rao's cousin's son, Damodar Rao, after the death of their son. However, it is believed that her husband, the Raja, never healed from the death of his son and died grieving for him on November 21, 1853. The British East India Company was gradually and assertively gaining control of several regions during this period. Jhansi provided access to the west and north and was in their plans. The East India Company exploited Maharaja's death by employing the Doctrine of Lapse. According to the policy, any princely kingdom under the direct or indirect sovereignty of the East India Company that did not have a valid male successor would be seized by the company. Moreover, no adopted son of the Indian emperor could be declared successor to the kingdom, contradicting and challenging the Indian ruler's authority to appoint an heir of their choice. Thus, Rao's rightful claim to the throne was rejected. Dalhousie then annexed Jhansi, saying that the throne had become "lapsed" and thus put Jhansi under his "protection". The Rani was awarded a pension of 60,000 rupees and asked her to leave the fort in March 1854. Rani Lakshmi Bai sought legal counsel from a British attorney. She subsequently filed an appeal, requesting that her case be heard in London. Her appeal, on the other hand, was denied. British authorities confiscated the state jewels. Furthermore, an order was issued for the Rani to abandon the Jhansi fort and relocate to the Rani Mahal in Jhansi.

The Sepoy Mutiny of India began in Meerut on May 10, 1857. This would serve as the impetus for the uprising against the British. It started when word got out that the new bullet casings for their Enfield rifles were smeared in pork/beef fat, although pigs are forbidden to Muslims and cows are revered by Hindus, rendering them forbidden to eat. British commanders insisted on their use, and anyone who did not comply was disciplined. Meanwhile, unrest spread throughout India, and the First War of Indian Independence began in diverse pockets across the northern subcontinent in May 1857. During this volatile period, the British were obliged to divert their attention elsewhere, leaving Lakshmi Bai to control Jhansi alone.

Her talents were continually demonstrated during this period, as she was able to lead her men effectively against riots erupting in Jhansi. Despite the Empire's turmoil, Lakshmi Bai's administration was able to maintain Jhansi relatively calm and quiet. Rani Jhansi was adamant about not abandoning Jhansi. She fortified its defences and gathered a volunteer army. Women were also trained in the military. Tantia Tope and Nana Sahib also backed her up.

On March 23, 1858, British troops arrived under Sir Hugh Rose and lay siege to Jhansi. Rani Jhansi and her devoted warriors made the decision not to surrender. During the two-week siege of Jhansi, Lakshmi Bai put up a valiant battle against the British. Jhansi was shelled mercilessly. Women were also carrying munitions and bringing meals to soldiers in the Jhansi Army. When the British army entered Jhansi after a violent battle, Rani Lakshmi Bai fastened her son Damodar Rao to her back and fought heroically with two swords in each hand.

She battled the British valiantly. An army led by the rebel leader Tantia Tope, was dispatched to relieve Jhansi and rescue Lakshmi Bai. However, the British, although having just 1,540 soldiers in the field to avoid breaking the siege, were better trained and disciplined than the recruits, and ultimately the untrained soldiers turned and fled quickly after the British began to attack. They were able to damage the city walls and seize the city three days later. Rani fled to Kalpi, accompanied by other companions. She then proceeded to Gwalior with Tantia Tope, where the combined rebel forces defeated the army of the Maharaja of Gwalior after his armies deserted the rebel forces. They then occupied the strategic fort at Gwalior.

She died on June 18, 1858, during the battle for Gwalior. Three days later, the British took Gwalior. General Sir Hugh Rose stated in his report on the struggle for Gwalior that the Rani, "remarkable for her beauty, cleverness and perseverance" had been "the most dangerous of all the rebel leaders."

It is believed her funeral was arranged on the same day near the spot where she was wounded. Her body was cremated at her request because she did not want the British to find it. A quick funeral was organised with the help of domestic help.

Her courage, tenacity and knowledge, as well as her progressive views on women's empowerment in nineteenth-century India and her sacrifices, made her an emblem of the Indian independence struggle. The Rani is remembered in bronze statues in Jhansi and Gwalior, which represent her riding a horse. Rani Lakshmi Bai rose to prominence as a national hero and was regarded as the ideal of feminine bravery in India. The Indian National Army's first female battalion was named after her. The Rani of Jhansi has been referred to as the "Joan of Arc" of the Indian independence movement.

Before asking others, one needs to set an example. The love of the queen towards the motherland, awakened the spirit of freedom among thousands of people. She inspired many women to participate in the freedom movement. She sacrificed her life for the dignity and freedom of her people. No doubt, she was a true queen and a courageous fighter.

KUYILI
(1700–1780)

5

KUYILI

An ignited soul who danced in the flames

Kuyili was an Army commander of Queen Velu Nachiyar in the 18th century, who fought in operations against the East India Company. Velu Nachiyar, the Queen of Sivagangai, was the first sovereign in Indian history to conduct war against British colonists in 1780. It is possibly less known that Kuyili, her Dalit Commander-in-Chief, executed the winning strategy. Kuyili played a significant role in molding India's history, although hardly ever acknowledged it. She is regarded as India's first suicide bomber and first female martyr. Kuyili was a dedicated commander of Velu Nachiyar. Some referred to Kuyili as Veerthalapathy (The Brave Commander) or Veeramangai (The Brave Woman). She was a Dalit valiant fighter from the Arunthathiyar caste.

Kuyili was born in Kudanchavadi, near the Sivagangai District, to farmers Periyamuthan and Raku. Her mother was equally renowned for her bravery, and she probably died fighting a wild bull to protect her fields. Periyamuthan moved near Sivagangai with Kuyili, where he worked as a cobbler. It is believed that he raised Kuyili by inspiring her with tales of her mother's many daring deeds. He was later hired as a spy by queen Velu Nachiyar, who was in hiding at the time. During the battle, he also fought with his daughter and the Queen. Due to the nature of Periyamuthan's occupation, Kuyili and Velu Nachiyar became close. Kuyili and her father were granted unlimited access to Velu Nachiyar at all times. Kuyili and Velu Nachiar established a special affinity. Kuyili once found that her Silambam (weaponised martial arts) instructor was a spy working against the Queen and rescued her from him. This led to the Queen making Kuyili her bodyguard. Later, Velu Nachiar appointed her as the Women's army's Commander-in-Chief.

To restore her empire, Velu Nachiyar went to war with the British after successfully forming alliances with the Maruthu Pandiyars, Hyder Ali and others. Even though their army was well-trained, they had challenges throughout the conflict due to the sophisticated armaments of the British Queen Velu Nachiar and Kuyili developed a plan to destroy the enemy's weapons. They were aware that only women would be permitted to enter the Sivagangaikottai (fort) because it was the tenth day of Navaratri and only women were permitted to celebrate the Vijayadashami festival. Kuyili commanded her force, which was disguised as civilians and surrounded the fort undetected by the British Army. The event was observed at the Rajarajeswari Amman shrine. Ammunition was stashed in the shrine, and Kuyili planned an ambush against the British. They concealed their weapons in flower and fruit baskets before entering the fort and attacking the British.

The British army was driven into battle after being taken by surprise. Kuyili, who had previously noted the location of the weapons' storage place, executed an alternative strategy. She took advantage of the disorderly environment and made her comrades pour ghee and oil used for lamp-burning over her. The woman then entered the storage room and lit herself ablaze. She destroyed all the weaponry, leaving the British force completely vulnerable.

The Government of Tamil Nadu dedicated a memorial to Kuyili in the district of Sivagangai. Kuyili's acts helped restore the identity of the Sivagangai people, although our history has failed to recognise her selflessness and bravery.

A woman who knows how to use her strength and wit for the sake of her people, Kuyili's supreme sacrifice for her nation, that was the world's first suicide bombing by any reckoning, remains an iconic act of unmitigated patriotism with a few parallel to it in India's colonial history.

UDA DEVI

6

UDA DEVI

Challenged the Convention and Strucked a Blow

Uda Devi, born in a small village in Lucknow, was one of the first Dalit women to fight the 1857 Indian uprising, also known as the Sepoy Mutiny. Her husband Makka Pasi was a soldier in the army of Hazrat Mahal, and upon learning of the British's threats to the country's independence, Uda Devi volunteered to fight against them. Both Uda Devi and her husband were part of the armed resistance when the British attacked Awadh. When she heard that her husband had died in the battle, she unleashed her final campaign in full force. During the fight of Sikandar Bagh, she is reported to have climbed a pipal tree and used a sniper rifle to kill several British soldiers. Ultimately, when the British started firing on the tree, she was also shot down. Devi and other female Dalit partners are today recognised as the warriors or "Dalit Veeranganas" of the 1857 Indian Rebellion.

In 1857, northern India was in a condition of near turmoil. A widespread revolution against the expanding authority of the British East India Company engulfed the towns of

Delhi, Jhansi and Kanpur. In Lucknow, a small British force clung flimsily to life at the Residency (a collection of buildings) on the river bank of the Gomti. Besieged by rebels and lacking basic supplies, the shrinking detachment of British soldiers, loyal Indian sepoys and non-combatants hovered on the edge of destruction throughout the summer of 1857. A relief column of British forces pushed through in September but lacked the weaponry to leave again.

In November, General Colin Campbell pushed through the rebel lines and finally saved the stranded garrison. During this "Second relief of Lucknow", Campbell's 93rd Highland Regiment pushed along the Southern bank of the Gomti river till they reached the Palace of Sikandrabad. The British forces met strong opposition from insurgents who had barricaded the location. It is said that in this battle, nearly 2000 rebels and many soldiers laid down their lives in hand-to-hand fighting.

Uda Devi dressed in men's clothing to join the rebellion, she was armed with a pair of heavy old-cavalry pistols, and kept a large amount of ammunition, while from her perch in the tree which was purposely prepared before the attack. After the British took over Sikandrabad, an officer observed that several of the British soldiers had bullet wounds indicating a steep, downward trajectory.

Deducing that a sniper remained hiding in a nearby pipal tree, British forces opened fire on the tree and evicted a rebel who fell dead to the ground. Subsequent research found that the rebel was Uda Devi Pasi, who had dressed in a man's attire to join the rebellion.

In the centre of a junction outside Sikandar Bagh in Lucknow, stands a statue brandishing a gun and walking forward with a resolute look. The plaque on the base of the statue described the figure as Uda Devi Pasi, a champion of the Mutiny of 1857. Today, Uda Devi is an excellent role model to women from non-dominant classes. Each year, on November 16, members of the Pasi caste assemble at the location of her final fall and commemorate her as an anti-imperialist rebel who broke convention and dedicated herself to the cause of Indian Independence. The legacy of Uda Devi's sacrifice is maintained alive by Dalit communities who respect her as a symbol of strong Dalit womanhood. They remember her as a fierce and courageous Dalit woman who opposed both caste and gender boundaries established in society.

Uda Devi is an inspiration to women and even men from non-dominant classes for her determination to serve her nation irrespective to the treatment vulnerable communities used to face those days. The class and audacity she had presented herself and shutting down the caste system should be acknowledged and hailed.

JHALKARI BAI

7

JHALKARI BAI

A woman who rose from hearth to battlefield

An extraordinary Dalit warrior and a female soldier named Jhalkari Bai made significant contributions to the Jhansi war. She played a significant role in the 1857 Indian Rebellion and served with Rani Lakshmi Bai of Jhansi's women's army. She later became an influential advisor to Rani of Jhansi, the queen. During the peak of the siege of Jhansi, she disguised herself as the queen and fought on the front lines on her behalf, helping the queen to flee the fort unharmed. According to the legends related to her, she was so fearless that she killed a tiger when it attacked her in the wilderness; therefore, her bravery is an intrinsic element of Bundeli culture.

Jhalkari Bai was born on November 22, 1830, in the village of Bhojla, close to Jhansi, and was the only daughter of Sadoba Singh and Jamuna Devi. Her family was of the caste Kori. Due to her family's poverty and belonging to the suppressed Dalit caste of Kori, she was unable to attend school and receive a formal education. Her father raised her when her mother passed away. She was taught to use weapons, ride a horse, and fight like a warrior at a very young age. During her youth, she reportedly stood her ground against a tiger and killed it with an axe. Pooran of Namapur Jhansi, himself of the Kori caste, heard her tales of heroism and daring since childhood. Pooram was a daring and renowned wrestler who was proficient in archery, horseback riding, firearms, and sword-wielding. According to rumors, he was a soldier of exceptional aptitude, and the court's generals quickly recognised his talents. He informed his mother of his desire to marry Jhalkari. In 1843, the marriage was solemnized with the approval of Jhalkari Bai's father. During Gauri Puja, Jhalkari accompanied several local women to the fort. There, the famed "Jhansi ki Rani" discovered her. She was shocked by the remarkable similarity she shared with Jhalkari and quickly asked about her. As soon as the queen was informed of Jhalkari's courageous deeds, she joined the women's wing of the army. Here, she was trained to fire and flare cannons as the army prepared for an invasion by the British.

In 1858, Field Marshal Hugh Henry Rose, 1st Baron Strathnairn, struck Jhansi to permanently put an end to the rebellion. As has been vividly documented in subsequent history books, the courageous queen engaged the British forces from her fort with only 4,000 soldiers. She could have held on to the fort for a long time, but she was betrayed by one of her commanders. Dulha Ju, who was in charge of one of the fort's gates, formed a bargain with the attackers and opened the doors of Jhansi to the British soldiers, making defeat inevitable. Rani Lakshmi Bai, advised by her generals to flee quickly, queen silently rode away from Jhansi with her son and attendants via the Bhandari gate to Kalpi. Jhalkari's husband Pooran was killed defending the fort but instead of mourning her loss, she worked out a plan to deceive the British.

When Jhalkari Bai learned of Rani Lakshmi Bai's escape, she immediately took action. Without a second thought for the danger, she was placing herself in, she dressed up as the queen and returned to the battlefield, where she assumed leadership of the army and approached General Rose's tent. This caused uncertainty that lasted a whole day, giving Rani's forces a restored edge. She cried loudly upon arrival that she desired a meeting with the General himself. This strategy was intended to both hold the invading forces at bay and give the queen ample time to dodge capture and escape danger. She engaged in combat between Dantiya gate, Bhandari gate, and Unnao gate.

Given her likeness to the queen, the British army was effectively duped for at least a day. According to Bundelkhand folklore, when the General (mistakenly believing she was the queen) asked her what penalty she deserved, she looked him in the eye and demanded to be hung. He is then said to have replied with admiration that if even one per cent of women in India were like her, then the British would have to leave the country defeated. The British did not discover the lie until the following day when a man named Dulha Ju identified Jhalkari.

Numerous Koli organisations commemorate the anniversary of Jhalkari Bai's passing as Shahid Diwas (Martyr Day). The attempt to establish Bundelkhand as a separate state also embraced the mythology of Jhalkari Bai to build the Bundeli identity. The Post and Telegraph department of the Indian government in 2001 released a commemorative stamp portraying Jhalkari Bai. The city of Gwalior has dedicated a statue to honour a warrior who died defending her people and her country.

It is mentioned in the 1951 novel Jhansi ki Rani by B.L. Varma, who incorporated a subplot concerning Jhalkari Bai. He referred to Jhalkari Bai as Koli, a distinguished soldier in Lakshmi Bai's army. In 1964, Bhawani Shankar Visharad penned the first biography of Jhalkari Bai.

She is free in her wildness, she is spontaneous. She knows death is inevitable, but courageous woman, who disguised herself as the queen and fought to let the queen escape safely out of the fort.

ANNIE WOOD BESANT
(1847–1933)

8

ANNIE WOOD BESANT

An indispensable soul for the Indian independence

In her defiance of social conventions, being an atheist, discontent with her role as merely a wife and a mother and later being separated from her clergyman husband, Annie Besant had amassed enough flak in the early part of her life to last her for a long time to come. And, as if all of this wasn't controversial enough at that time, Besant advocated for women's rights and birth control by republishing Charles Knowlton's Fruits of Philosophy, a book that was a manual on birth control and had advanced ideas for the time. As biographer Rosemary Dinnage noted, this was a lady with a "powerful will." For all the battles she fought and endured all her life, Besant was courageous and had unwavering conviction. A woman of Irish descent, she became the first President of the Indian National Congress, chairing the 1917 session of the Indian National Congress in Calcutta (now Kolkata).

Anne Wood Besant was a famous British socialist, women's rights activist who advocated for women's access to birth control and marriage reform, theosophist, orator, writer and supporter of Irish and Indian self-rule and supporter of Indian nationalism. Annie Besant was born on October 1, 1847, to Emily Morris and William Wood in London, UK. She lost her father when she was just five years old. Annie had an unhappy childhood exacerbated by the death of her father. Interestingly, Anne's initial life was immersed in religion. Her mother worked hard in the countryside and ran a boys' boarding house at Harrow School to support her family. Her mother could not support her and persuaded her friend Ellen Marryat, the sister of the writer Frederick Marryat, to take responsibility for her daughter. Marryat was a prosperous Harrow citizen who established a private school for several needy children and ensured Besant received a quality education.

At age 20, Annie married Frank Besant, a 26-year-old clergyman, and they had two children. After a year of marriage, Besant met William Prowting Roberts, a family friend and lawyer who frequently represented impoverished women and children in the Labour Movement. The Labour Movement mostly comprised of socialists who fought for better working conditions for the poor. Besant learnt a lot from her frequent visits to Roberts' office, where she observed his work and saw him push for labour unions and better working conditions. Her support for workers and weakening faith in religion were sources of tension in her marriage. She soon separated from her husband in 1873. She always fought for what she believed to be right. Though she did not lose her mother's pride and remained in touch with her children.

In 1878, Besant assisted in the establishment of the Malthusian League and acted as its secretary. The Malthusian League was founded on Thomas Malthus's writings, in which Malthus asserted that overpopulation was imminent since humans' reproductive capacity exceeded the earth's capacity to maintain an increasing population. In the 1870s, the Malthusian League argued for access to contraception and family planning as a means of avoiding overpopulation and poverty. Besant was engaged in both the Malthusian League and other socialist organisations, and in 1878, she was elected president of the socialist reform group, the International Labour union. Besant was enrolled at the University of London in 1879 and studied botany and physiology there. She was one of the first female students admitted to college.

Besant backed a series of worker demonstrations calling for improved working conditions. She assisted in the organisation of a strike by female workers at the Bryant and May match factory in east London in 1888. She published "White Slavery in London," documenting the horrible working conditions endured by the women match workers and the factory's phosphorous fumes having a detrimental effect on their health. She distributed this piece among women, along with roses. This garnered immediate support among revolutionaries as well as widespread press coverage, as a result their managers eventually improved their working conditions significantly.

It was as if fighting for social and political reforms and being surrounded by controversies were insufficient for Besant. She developed an interest in Theosophy, a religious movement founded in 1875. She was a gifted writer and an eloquent speaker. She was invited in 1889 to write reviews for the Pall Mall Gazette on The Secret Doctrine (a book by the Russian-born religious mystic Helena Blavatsky, a co-founder of the Theosophical Society). She became a Theosophist following contact with Madame Blavatsky and dropped her support for the Knowlton pamphlet she had previously championed with such zeal. Theosophists believed that spiritual relationships and intuition could be used to attain enlightenment and salvation. However, it was theosophy that ultimately brought her to India. Besant became a member of the Theosophical Society, which was founded in New York City and eventually headquartered in Chennai, India. She made her first visit to Chennai, India in 1893 to attend the Theosophical Society's Annual Convention in Madras. Five years later, she founded the Central Hindu College in Benares (now Varanasi).

Annie Besant advocated for Indian independence, which was then under British authority. Besant was moved by the intense conflict for independence from British domination and gradually became an active participant. Her most significant contribution to the Indian freedom struggle was the establishment of the Home Rule League in 1916 together with Lokmanya Bal Gangadhar Tilak. They continued the historic campaign that became a watershed moment in India's decades-long quest for independence. The crusade's objective was to secure for India the same Dominion status as Australia and Canada. The campaign lasted two years, with the Indian Home Rule League's operations assisting the liberation struggle significantly. Besant was arrested in the summer of 1917 for publicly opposing the British colonial authorities. She was released after nearly four months and after the Indian National Congress and the Muslim League threatened to begin protests against the British administration. Additionally, she founded the Young Men's Indian Association to educate and encourage people for public service, as well as the daily newspaper New India, which became the primary vehicle through which she and her fellow members promoted Home Rule. In December 1917, she was elected President of the Indian National Congress.

Her nationalism caused immense discomfort to the colonial authorities, and her Home Rule campaign was labelled positively disloyal. With World War, I raging and Besant unflinching in her journalistic advocacy and public remarks, she was arrested and imprisoned in Ooty for a while. However, the Indian population regarded her as a hero. Annie Besant was a courageous woman who was called a freedom warrior due to her involvement in numerous wars to assist people in reclaiming their rights and liberty. She was an active participant in the Indian independence movement and pushed for India's freedom. As a writer and speaker, she admired and understood the Indian people, culture and traditions. In 1893, she declared India to be her own country and began rousing Indians from their profound sleep with her powerful speeches. Mahatma Gandhi once said that she roused the Indians from their slumber.

For nearly twenty-five years, she served as the president of the Theosophical Society in India, until she died in 1933. Besant remained unwell for several years after falling down the stairs at the Theosophical Society headquarters. She died in Chennai in India on 20th September 1933. This political ground breaker achieved such prominence in India that in 1917 she was granted the title of 'mother.'

It takes a lot of heart to devote their purpose to a land where they weren't even born. Annie Besant put her whole heart and soul together in all that she embarked on, she teaches us that the geographical boundaries should not be an obstacle if you are determined to do something extraordinary for the welfare of the people.

BHIKAJI RUSTOM CAMA
(1861–1936)

9

BHIKAJI RUSTOM CAMA

"I am not free, until anyone is unfree"

Madam Bhikaji Cama became the first person to unfurl the Indian Flag on foreign land in Stuttgart, Germany, on August 22, 1907. It was International Socialist Conference that drew 1,000 delegates from all over the world. "Behold, the Flag of independent India is born! It has been made sacred by the blood of young Indians who sacrificed their lives in its honour. In the name of this Flag, I appeal to lovers of freedom all over the world to support this struggle." These words were spoken by fiery Bhikaji Cama. She boldly stated: "This is the Flag of independent India. I appeal to all gentlemen to stand and salute the Flag. Madam Bhikaji Cama also highlighted the devastation caused by a famine that had ravaged the Indian subcontinent, pleading for human rights, equality, and independence from Great Britain. Madam Bhikaji Cama's determination, courage, and honesty gave the Indian liberation cause a new level of intensity and clarity. The fierce lady had once said, 'Do not forget the important role of women play in building a nation.'

A prominent personality of the Indian Nationalist Movement, she was born as Bhikaiji Patel on 24th September

1861, into a very prosperous Parsi business family in Bombay (now Mumbai) and obtained her early education in Bombay. Her parents, Sorabji Framji Patel and Jaijibai Sorabji Patel were well-known in the city. Sorabji was a lawyer by training and a merchant by profession and was a prominent name in business, education and philanthropy in Bombay. She was drawn to political concerns at a young age as a result of growing up in an environment where the Indian nationalist movement was coming to the fore. She married Rustomji Cama, a well-known lawyer and a pro-British in 1885, but her participation in sociopolitical concerns caused differences in the marriage. While Mr Cama admired the British, respected their culture, and believed they had done much good for India, Bhikaji was a true nationalist who believed the British had mercilessly manipulated India to show their power and grow.

In 1896, when Bombay was struck by a succession of health tragedies, including hunger and subsequently bubonic plague, Bhikaji instantly offered to assist the several teams tasked with saving plague victims. She got the plague too while she cared for the sick, but survived. Due to her debilitated state, she was advised to travel to Europe for rehabilitation. In 1902, Bhikaji left India for London, due to bad health that necessitated medical attention, and also due to marital difficulties. Europe would become her home for the remainder of her years.

In 1904, she met Shyamji Krishna Varma, a well-known figure in London's Indian circle for his ardent patriotic speeches delivered in Hyde Park. She also met Dadabhai Naoroji, a staunch critic of British economic policy in India, during her stay she began working for the Indian National

Congress. Cama also came in contact with Indian nationalists such as Veer Savarkar and Lala Har Dayal. They quickly realised that peaceful protests would not suffice to end British rule. Madam Cama and other revolutionaries began giving talks in Hyde Park. These impassioned speeches drew the attention of local Indians living abroad as well as foreign journalists. Shyamji Verma established the India House in London, which became a nexus for Indian revolutionaries living in the UK. The India House was founded on July 1, 1905. The inaugural ceremony was attended by revolutionaries such as Lala Lajpat Rai and Dada Bhai Naoroji. Madam Cama used to meet at India House with other revolutionaries like Veer Savarkar and Senapati Bapat to plot the demise of British rule.

Lala Lajpat Rai was captured in India during the Golden Jubilee of India's first fight for independence. At the time, Madam Cama wrote vehemently in The Indian Sociologist, pleading with readers to continue their non-cooperation. She saw that the most effective method of dismantling Britishers in India was to stop cooperating with them; this would quickly bring their regime to a halt. Veer Savarkar, with the backing of Madam Cama, became the leader of The India House and founded a group called 'Abhinav Bharat'. This group was used to commemorate Indian freedom fighters' birthdays. When English authorities banned Veer Savarkar's book 'First War of Indian Independence' before it could be published, it was Madam Cama who got it smuggled into India and translated it into French.

In 1907, she attended the International Socialist Congress at Stuttgart. She holds the distinction of unveiling the original form of India's national Flag—In the Flag, the top green stripe had eight blooming lotuses representing pre-independence India's eight provinces. 'Vande-Mataram' was written across the central saffron stripe in Hindi. On the bottom red stripe, a half moon was on the right and the rising sun on the left, indicating the Hindu and Muslim faith. The Flag which she unfurled at Stuttgart was smuggled into India by other independence activists. It is displayed at the Maratha and Kesari Library in Pune to this day. The independence banner Flag served as the prototype for India's present national Flag.

As a result of the spectacular episode, all the conference's representatives arose and saluted the first Flag of independent India. Madam Cama desired to bring to the international community's notice the poverty, famine and persecution endured during the British Raj, as well as India's desire for liberty. She was successful in this major feat. The world was unaware of the fervent patriotism of hundreds of young Indians willing to die to liberate their nation from colonial control. Additionally, the British attempted to punish revolutionaries at the time by enacting laws, imposing restrictions, and locking up them for life. Madam Cama's conduct and remarks constituted treason against the British Raj. Such was her fearlessness! Had she been apprehended; she would have been exiled to the infamous Kala Pani jail in Andaman.

Cama travelled on an extended lecture tour following the 1907 Stuttgart conference to mobilise public mood against British rule in India, particularly among expatriate Indians; she also advocated for women's rights. When speculations of her deportation from England began, she relocated to Paris in 1909. There she co-founded the Paris Indian Society with Singh Rewabhai Rana and Munchershah Burjorji Godrej. Her house became a headquarters for those fighting for Indian independence. Cama, along with other prominent exiled members of the Indian nationalist movement, authored, published and disseminated revolutionary writing for the movement, including Bande Mataram and, subsequently, Madan's Talwar (in response to the execution of Madan Lal Dhingra). Bhikaji managed to send the weekly periodicals to Indian revolutionaries despite being banned in India and Britain. Her writings made their way to India through the French colony of Pondicherry (now Pudducherry). She also assisted the revolutionaries in whatever way she could, whether it was with money, materials or ideas. Despite their greatest efforts, the British government was unable to restrain her.

Bhikhaiji Cama was a fervent supporter of gender equality, influenced by Christabel Pankhurst and the Suffragette movement. In 1910, when speaking in Cairo, Egypt, she inquired, "I see here the representatives of only half the population of Egypt. May I ask where is the other half? Sons of Egypt, where are the daughters of Egypt? Where are your mothers and sisters? Your wives and daughters? However, her stance on Indian independence was far greater than her position on any other issue. It is said that when Cama met Herabai and Mithan Tata in 1920, two Parsi women who were vocal on the matter of the right to vote, she sadly said, "Work for India's freedom and independence. When India is independent women will not only have the right to vote but all other rights."

Following the assassination of William Hutt Curzon Wyllie in 1909, the authorities in London began clamping down on Indian nationalists residing there. Vinayak Damodar Savarkar was arrested together with a large number of other activists following Wyllie's assassination in 1909. The following year, while being taken by ship back to India for trial, he managed to flee when the ship was docked in Marseilles harbor. He was unable to locate Bhikaiji Cama and others who were expecting him and fell into British custody once more. French authorities declined to cooperate with the British authorities in extraditing Bhikaiji Cama. Following this, the British government seized her inheritance. Lenin reportedly extended an invitation to her to live in the Soviet Union, but she declined.

The situation became complicated when France and Britain became allies during World War I in 1914, and all the members of the Paris India Society except Cama and Singh Rewabhai Rana left the country. In January 1915, the French exiled Rana and his entire family to the Caribbean Island of Martinique, while Cama was detained in Vichy. She was released in November 1917 in poor health and permitted to return to Bordeaux on the condition that she report weekly to the local police. Cama returned to her house in Paris following the war. Her relationships with Indian and foreign revolutionaries were unaffected, and she maintained contact with the movement.

Madam Cama lived in exile for the majority of her life, in foreign nations. Cama stayed in exile in Europe till 1935, when, critically ill and paralyzed following a stroke earlier that year, she appealed to the British authorities for permission to return home through Sir Cowasji Jehangir, a prominent member of the Bombay Parsi community. On 24 June 1935, she eventually agreed to the condition that she abstain from seditionist actions in a letter from Paris. She arrived in Bombay in November 1935, escorted by Jehangir, and died nine months later, on 13 August 1936, at the age of 74.

Bikhaiji Cama bequeathed a sizable portion of her wealth to the Avabai Petit Orphanage for girls, now the Bai Avabai Framji Petit Girls' High School, which established a trust in her name. She also donated substantially to her family's fire temple, the Framji Nusserwanjee Patel Agiary at Mazgaon, in South Bombay.

Numerous streets and sites in India are named after her in recognition of her efforts and contribution. On the 11th Republic Day of India in 1962, the Indian Posts and Telegraphs Department produced a memorial stamp in her remembrance. In 1997, the Indian Coast Guard commissioned a Priyadarshini-class fast patrol vessel ICGS Bikhaiji Cama after Bikhaiji Cama's name.

Despite living in exile for the majority of her life, she worked tirelessly for India's independence. While she did not live to see the Flag she raised on that momentous day in Germany, she fought tirelessly to ensure that it would one day unfurl for all of us to be proud of.

Bhikaji Rustam Cama is someone who knows to turn the obstacles into her stepping stones. She was a persistent and a tenacious woman with a strong indomitable trait having her own identity and trajectory that she followed. Her ambitions and goals were synced to the nation's freedom.

KASTURBA GANDHI
(1869–1944)

10

KASTURBA GANDHI

The woman deeper than the roots of trees

Kasturba was first imprisoned in Durban in 1913 for defying the Cape Supreme Court's decision. She was a courageous woman, with unflinching will and immense patience, even though her life was too entwined with that of her husband, Mohandas Karamchand Gandhi. Despite being known as the wife of a renowned man, Kasturba was a passionate patriot who tirelessly worked for civil rights and India's independence from British rule. Undoubtedly, Kasturba was devoted to her husband and was resolutely by his side no matter what. Her patience is said to have inspired Gandhi's call for Satyagraha.

Kasturba was born into the wealthy and illustrious Kapadia family of Porbandar, Gujarat. Her father was a wealthy trader named Gokuladas Kapadia, and her mother's name was Vrajkunwerba. Kasturba was born in Porbandar on April 11, 1869. She received no formal schooling, as was customary in traditional families during this era. In 1882, she was married to Mohandas, the younger son of Karamchand Gandhi, Porbander's dewan. They had four sons. The husband taught his wife to read and write in Gujarati, their native tongue, and she picked up enough language to read daily newspapers. Despite being an extraordinary woman with an independent spirit, her work was usually eclipsed by that of her more famous husband. Needless to mention, it was his wife, Kasturba Gandhi (1869-1944), whom Gandhi mourned upon her death after more than sixty years of marriage: "I can't imagine a life without Ba. She went away to freedom, imprinting on the heart to work or to die."

Gandhi, the advocate of ahimsa (non-violence) and satyagraha (non-violent resistance), expressed on various occasions that Kasturba taught him the art and science of satyagraha. Her life was defined by love, loyalty, selflessness and silence. Gandhi stated in his book that she had assisted him, voluntarily or otherwise, in all of his life's major shifts. He regarded her life as incredibly holy. She never stood in the way of his self-sacrifices. Everyone referred to her as Ba with tremendous reverence. She too was influenced by her husband's views and soon became involved in political activism herself.

Meanwhile, her spouse was gaining popularity as a result of his political engagement. She, too, collaborated closely with her husband and travelled with him to South Africa in 1897 when he went to practice law. This journey to South Africa brought Kasturba to the forefront. During the 1910s, she became deeply committed to campaigning at the Phoenix Settlement near Durban. They lived in Durban for more than 19 years. At the same time, she juggled her responsibilities as a mother. She was imprisoned and condemned to three months in a hard labour prison for her participation in the 1913 protest against South African Indians' working conditions, but her views never wavered. Kasturba inspired women in India to join the Indian independence struggle and encouraged them to volunteer for socio-political causes. In 1906, Mohandas swore an oath of celibacy and chastity and Kasturba backed her husband willingly. Mohandas returned to India in 1915 to support the Indian independence cause, and his wife, always encouraging, accompanied him. Furthermore, she became interested in social work, instructing children in reading, writing, health and hygiene. When Gandhi established the Satyagraha Ashram at Sabarmati, Ba served as his chief assistant in the ashram's administration. Ba organised the propaganda for Gandhi's khadi movement. Later, she worked for the empowerment of indigenous workers to make 'Swadeshi items.'

Whenever Gandhi embarked on a fast against what he perceived to be unjust, Kasturba accompanied him. She led the women's satyagraha campaign and was imprisoned several times. As her husband became more engaged in political causes and protests, she joined him and became a pillar of support for him. Kasturba also assisted Gandhi in 1917 in improving the conditions of women in Champaran, where he was working for the welfare of indigo planters. She took part in non-violent protests against the British in Rajkot in early 1939. Gandhi's regular fasting also took a toll on her health. She was arrested and held in solitary prison for a month, which further deteriorated her health. In 1942, she was arrested for her involvement in the Quit India Movement. She experienced two heart attacks in January 1944, and never fully recovered.

Kasturba Gandhi who was the shadow of her husband who became the father of the nation died on 22 February 1944. Mahatma Gandhi's leadership was partly his own, but he also owed some part of it to his wife. All said and done, he was self-aware and honest enough to acknowledge his appreciation for her.

"Ba" that is truly mother to us. A woman who bore the burden so cheerfully, so steadfastly, so triumphantly throughout her husband's stormy sea of life for over 60 years. The struggle, aspirations and hard work of an ordinary simple women are remarkable who transmitted Mohandas Gandhi to Mahatma Gandhi, the Father of Nation.

MATANGINI HAZRA
(1870–1942)

11

MATANGINI HAZRA

An old soul with young eyes

Carrying the Indian Flag, the 73-year-old stepped into a hail of bullets. In September 1942, 73-year-old Hazra led a parade of over 6,000 demonstrators, most of whom were women. The procession marched intending to seize control of the Tamluk police station from the British. During the confrontation that ensued between demonstrators and police, the British police officers shot her three times. She marched on, despite being wounded, with the Flag, singing 'Vande-Mataram,' until she collapsed and died clutching the Flag in her hands. The Tamluk woman was a devout disciple of Gandhi and was imprisoned numerous times. She was affectionately known as Gandhi Buri, which means "old lady Gandhi" in Bangla.

Matangini Hazra was an extraordinary woman for her time. She was born in 1869 in a village called Hogla, near Tamluk in Bengal Presidency. Although little information about her early years is publicly available, research conducted in the West Bengal government archives indicates that she was the daughter of a poor farmer who could not afford to provide her with formal education. She was married to Trilochan Hazra, a resident of Medinipur's (or Midnapore) Alinan village. Matangini Hazra was widowed and childless after few years of her marriage.

The years passed and she began dedicating her time to social causes. The Nationalist movement gained popularity across the subcontinent in the early 1900s, and Gandhi travelled widely throughout the region, raising awareness for the freedom struggle. She was inspired by Gandhi's philosophy and became highly involved in the Indian independence movement. Women's engagement in the Midnapore liberation struggle was remarkable during this period.

However, a few years later, a defining moment occurred in her life. On 26th January 1932, men from the neighborhood marched in a parade to raise awareness about the country's political situation. She came out and joined them as it went past her cottage. At that moment, Matangini vowed to fight for India's independence from the British.

She was jailed in 1930 at the age of 61 for participating in Civil Disobedience Movement and the Salt March. She was quickly released, but she insisted on the abolition of the tax. During a march towards the court building, to protest against the Governor's severe decision to punish all agitators, Matangini was detained again and imprisoned in Baharampur for six months. Her involvement in the Civil Disobedience Movement resulted in many brief prison sentences, during which she met other female revolutionaries who had also been imprisoned for anti-British actions. Following her release, she joined the Indian National

Congress and began spinning her own Khadi, despite her failing eyesight and advancing years of age. In 1933, she attended the sub-divisional Congress conference at Serampore and was injured in the ensuing baton charge by the police. Her sense of service and dedication was such that she tirelessly nursed the sick men, women and children following the advent of a smallpox epidemic.

Hazra remained staunchly committed to the cause of independence from British domination. Her role in the freedom struggle grew stronger during the August 1942 Quit India Movement. Members of the Indian National Congress's local branch in Medinipur proposed seizing control of the district's local government buildings and police stations as the first step towards attaining their goals in terms of the Quit India Movement. In September of that year, a 73-year-old Hazra led a parade of over 6,000 demonstrators, most of whom were women. The procession marched to seize control of the Tamluk police station from the British. The Crown Police ordered the march to dissolve as it reached the outskirts of town, citing Section 144 of the Indian Penal Code.

During the confrontation that ensued between demonstrators and police, Hazra approached the police with the freedom movement's Flag in hand, pleading with them not to shoot at the procession. Her cries for help were ignored, and British police officers shot her three times. Wounded and unperturbed, she marched with the freedom Flag, singing 'Vande-Mataram,' until she collapsed and died clutching the Flag in her hands. The parallel Tamluk administration incited open revolt by honouring her "martyrdom for her country" and functioned two more years before being terminated at Gandhi's insistence in 1944.

Matangini Hazra was the first woman revolutionary to have a statue dedicated to her at the Kolkata Maidan in 1977. In her home town, Tamluk, a similar statue commemorates the location where she was killed by British officials. The postal service released a series of postage stamps in 2002 to commemorate sixty years of the Quit India Movement. Among them was a five-rupee postal stamp featuring a replica of Matangini Hazra. A key road in south Kolkata, Hazra Road, is also named after her. These honours serve as fitting reminders of her extraordinary life. She remained away from public attention, but her efforts were priceless.

The age when, we think we are retired and have been titled as senior citizen, Matangini Hazra was fighting for the Nation's Independence and was behind bars for it at 61 years of age. Not stopping here, at the age of 73, she continued to lead processions, while being continuously chased by the Police. She died for motherland showcasing her selfless valor act.

BIBI GULAB KAUR
(1890–1941)

12

BIBI GULAB KAUR

Instincts over upbringing

Bibi Gulab Kaur was leader of Ghadar Party in Manila, Philippines. She and her husband Mann Singh sailed from the Philippines to India in order to rally the masses against British tyranny. Besides spreading anti-British public sentiment, she distributed literature relating to freedom movement, arms and ammunition to the members of Ghadar Party. Under sedition charges, she was kept in the notorious Lahore Fort and tortured for two years.

Gulab Kaur was born in 1890 in the village of Bakshiwala in the Sangrur Region of Punjab in a poor peasant family. She was married to Mann Singh. This was a time when many Punjabi farmers were expelled from their land by British policy. The couple departed for Manila, Philippines with the intention of emigrating to the United States of America in order to escape economic troubles in their own country. In Manila, she came in contact with Baba Hafiz Abdullah (Fajja), Baba Banta Singh, and Baba Harnam Singh (Tundilat), founders of the renowned Ghadar Party, an Indian revolutionary organisation created mostly by Punjabi Sikh immigrants' intent on removing the British from India.

Gulab Kaur joined the fight to oust the British in India and became one of the key Ghadar party members. She travelled around the Philippines in order to organise the Indian citizens. Moreover, she gathered weapons and funds for the party. She mobilised people in Manila and inspired them to join the party by distributing independence literature and delivering inspiring speeches to Indian passengers of ships.

Although some versions indicate that the couple intended to return to India to fight the British, Mann Singh ultimately chose to go to the United States. Meanwhile, Gulab Kaur chose to leave her husband and join the Ghadar movement without looking back. She returned to India in response to the Ghadar party's call and continued her revolutionary activities in Punjab. As soon as she arrived in India, she became an active ally in locations like Kapurthala, Hoshiarpur, and Jalandhar, motivating the public for armed revolution for the cause of the country's independence. One of the things she performed closely was to disseminate literature associated with the independence struggle, and keeping a close eye on their revolutionary printing machine.

In addition to promoting anti-British sentiment, she delivered weaponry and ammunition to Ghadar Party members while posing as a journalist. She also encouraged others to join the Ghadar Party. Unfortunately, the British authorities caught up with her and she was sentenced to two years in prison on charges of sedition in Lahore. She was held captive and tormented for two years in the infamous Lahore fort. But she continued to rally people against the British and their supporters in spite of this.

She passed away in 1941. There is little academic research on Gulab Kaur. S Kesar Singh has published a Punjabi book named *Gadhar di dhee-Gulab Kaur* (Gadar's Daughter-Gulab Kaur), but little else is known about it.

Bibi Gulab Kaur knew that the great cause involves high risks. Those were the days women used to sit behind curtains and listen to speeches, but not Bibi Gulab Kaur. She volunteers to become present in all the struggles and risked her life to get safety and freedom for all of us.

SAROJINI NAIDU
(1879–1949)

13

SAROJINI NAIDU

India's Nightingale

Sarojini Naidu had little expectation of becoming a poetess and being called 'the Nightingale of India' by then-President Rajendra Prasad when she began writing at a young age. Her poetry found its way into her patriotic speeches and Constituent Assembly debates. Naidu, an ardent member of the freedom movement, delivered passionate speeches across the nation. She discussed women's rights and labour rights and asked others to join the fight. She created the Women's India Association in 1917 with the assistance of Annie Besant and, like many others in the Constituent Assembly, concentrated on women's education and the abolition of child marriage. She was also the first woman governor of an Indian state.

Born in 1879 to a Bengali family with seven siblings, Sarojini completed high school at the age of 12. Her father was the principal of Hyderabad college and held a doctorate of Science from Edinburgh University. Her mother wrote poetry in Bengali. Sarojini was a highly bright and brilliant child. She spoke several languages fluently, including

English, Bengali, Urdu, Telugu and Persian. Young Sarojini was drawn to poetry, despite her father's wishes for her to become a mathematician or scientist.

She utilised her remarkable writing skills to compose a 1300-line poem named "The Lady of the Lake" in English. Dr Chattopadhyaya encouraged Sarojini's writings because he was impressed by her ability to convey emotions with the right words. A few months later, with the support of her father, Sarojini composed the Persian-language play "Maher Muneer." Sarojini's father distributed some copies of the play among his friends and relatives. He also sent a copy to the Nizam of Hyderabad. Impressed with the works of the little child, the Nizam granted her a scholarship to study overseas. At the age of 16, she got admission to the King's College, England and later joined Girton College in Cambridge. She also met Gopal Krishna Gokhale in England, who inspired her to join the independence movement.

Sarojini Naidu lived an active literary life. Her first collection of poetry, The Golden Threshold, was published in London in 1905. Edmund Gosse proposed the publishing, which featured a preface by Arthur Symons. John Butler Yeats also sketched Naidu as a teenager wearing a ruffled white dress. She had earlier read The Gift of India, a poem criticizing the British empire's abuse of Indian mothers, at the Hyderabad Ladies' War Relief Association in 1915. In 1928, a collection of all of her published poems was published in New York. The Feather of the Dawn (1961), edited by her daughter Padmaja Naidu, is a collection of Naidu's complete poems, including unpublished pieces, compiled after her death. The talks of Sarojini Naidu were initially compiled and published in January 1918 as The Speeches and Writings of Sarojini Naidu.

In 1914, Sarojini first visited Mahatma Gandhi at his residence in Kensington, England. At the request of her lifelong mentor, Gopal Krishna Gokhale, she met several famous people associated with the Indian freedom struggle. She met Mahatma Gandhi, Pandit Jawaharlal Nehru, C.P. Ramaswami Iyer, and Muhammad Ali Jinnah. Her bond with Gandhi was characterised by mutual respect and good humor.

In 1905, during the Partition of Bengal, Naidu joined the Indian National Movement after being inspired by the two individuals. She became involved in the Indian nationalist movement and a devotee of Mahatma Gandhi's concept of swaraj. She met Jawaharlal Nehru in 1916, worked alongside him to improve the depressing conditions of the Indigo workers of Champaran in the western province of Bihar, and campaigned fiercely for their rights against the British. Sarojini Naidu travelled around India and gave speeches on youth welfare, labour dignity, women's freedom, and nationalism. The years between 1917 and 1919 were the most turbulent in Sarojini's career. She advocated for the Montagu Chelmsford Reforms, the Khilafat problem, the oppressive Rowlett Act, and the Satyagraha at this time. When Gandhi initiated the Civil Disobedience Movement, she was a devoted lieutenant. She subdued the rioters with tremendous bravery, sold banned publications, and spoke at frenzied meetings about the massacre at Jallianwala Bagh in Amritsar. In 1917, she helped Annie Besant and other notable activists to create the Women's India Association. She also addressed to Congress the need for increased female participation in the independence movement. As the Flagbearer for the Indian Nationalist movement, she travelled widely throughout the United States and many European nations. She was arrested, along with other leaders such as Mahatma Gandhi, Jawaharlal Nehru, and Madan Mohan Malaviya, for taking part in the 1930 Salt March.

In March 1919, the British government introduced the Rowlatt Act, which made it unlawful to possess seditious documents. To protest, Mahatma Gandhi organised the Non-Cooperation Movement, and Naidu was the first to participate. Sarojini Naidu followed Gandhi's example religiously and actively supported his subsequent activities, including the Montagu-Chelmsford Reforms, the Khilafat problem, the Sabarmati Pact, the Satyagraha Pledge, and the Civil Disobedience Movement. When Gandhi was jailed in 1930 following the Salt March to Dandi, she and other leaders conducted the Dharasana Satyagraha. In 1931, she joined Gandhi in London for Round Table Talks with the British government. Her political efforts and participation in the Freedom struggle landed her in prison multiple times, in 1930, 1932, and 1942. Naidu succumbed to a heart arrest on March 2, 1949.

Sarojini Naidu was the first women Governor of Uttar Pradesh from August 1947 to March 1949. She was awarded the Kaiser-i-Hind medal by the British government for her vehement work during the plague epidemic in India. The name "Kaiser-i-Hind" literally means "emperor of India" Sarojni has eloquently captured the hearts and emotions of Indians with her poetic treasure. She aimed to compose largely brief poems, each addressing a specific idea or emotion. She has addressed a variety of topics, including Indian culture, the beauty of nature, the philosophy of life, Indian women, and patriotism. Several institutions, including the Sarojini Naidu College for Women, Kolkata, Sarojini Devi Eye Hospital, Hyderabad and Sarojini Naidu School of Arts and Communication in the University of Hyderabad are named in her honour.

From Sarojini Naidu's life we learn to break old stereotypes, spread our wings in multi directions, against the wind, fly high, and believe that sky is not the limit. She was a multifaceted personality, and possessed a great combination of empathy and audaciousness.

BASANTI DEVI
(1880–1974)

14

BASANTI DEVI

The adopted Mother

Basanti Devi participated in the liberation movement and was arrested during the 1921 non-cooperation movement. She served as president of the Bengal Congress from 1921 to 1922. After her husband, C R Das arrest, she assumed responsibility for the publication Bangaler Katha. It was she who pushed Indian youth to assert Indian dignity after the British police assault on Lala Lajpat Rai. She participated in the Khilafat and Civil Disobedience movements. In addition, she was one of the founding members of the Nari Karma Mandira, a women's education center.

Basanti Devi was born on the 23rd of March, 1880. During British colonial authority, her father Baradanath Haldar was the diwan of a large-scale zamindary in Assam. Basanti Devi studied in the Kolkata Loreto House. Here, she met her spouse Chitaranjan Das, and the couple got married.

Basanti Devi participated in numerous movements, including the Civil Disobedience Movement and the Khilafat Movement, after her husband passed away and carried out his missions. Destiny again played a tragic role and over took her with the death of only son after one year of her husband's death. She remained firm and strong as she had to look after her daughter-in-law along with her three daughters. Basanti Devi with her extreme valour subsequently participated in the 1920 session of the Indian National Congress at Nagpur. The following year, she collaborated with Urmila Devi and Sunita Devi, sisters of Das, to establish a training institution for women activists called Nari Karma Mandir. She was crucial in the collection of gold ornaments and 2,000 gold coins from Jalpaiguri for the Tilak Swaraj Fund during 1920–1921. The Indian National Congress called for strikes and a ban on foreign goods during the Non-cooperation agitation of 1921. Small groups of five volunteers were engaged in Kolkata to sell khadi, the hand-spun clothing, on the city's streets.

Basanti Devi took to the streets despite Subhash Chandra Bose's concerns that it would incite the British to arrest her. She was arrested for the same offence, as anticipated, but was released at midnight. The arrest caused tremendous unrest. Two prisons in Kolkata were packed with revolutionary volunteers, and hurriedly constructed detention camps were used to hold additional detainees. On December 10, 1921, the police apprehended Das and Bose. Basanti Devi assumed control of Das' weekly journal Bangaler Katha following his arrest.

She also served as president of the Bengal Provincial Congress in 1921 and 1922. Through her speech at the Chittagong meeting in April 1922, she supported grassroots activism. As she travelled around India, she advocated for the cultural advancement of artists that challenged colonialism. In 1928, Lala Lajpat Rai, an Indian freedom fighter, died after being hurt by police during a baton charge during his peaceful protest march. Basanti Devi then urged the Indian youth to revenge for the death of Lajpat Rai. As Das was Subhash Chandra Bose's political mentor, Bose had high admiration for Basanti Devi and considered her to be his "adopted mother". Like her husband, Basanti Devi was supportive of revolutionary activists in the Indian independence movement. In 1947, after India's independence, Basanti Devi maintained her social efforts. Basanti Devi College, the first government-funded women's college in Kolkata, was established in 1959 and named after her. She established the Chittaranjan Seva Sadan for the benefit of disadvantaged mothers. In her honour, the government constructed the Basanti Devi College in Kolkata. In 1973, she was also awarded the Padma Vibhushan, India's second-highest civilian decoration. She passed away in 1974 at the age of 94 years.

A woman who supported her husband, walked with him for a common cause of India's independence, and after him became the torch bearer, undaunted, unbroken in spirit. A woman with strong head and calm heart knows how to deal with all the odds.

UMABAI KUNDAPUR
(1892–1992)

15

UMABAI KUNDAPUR

Courageous under Crisis

Women's freedom was still constrained and controlled in those times. Umabai overcame a near-impossible mission when she mobilised over 150 women volunteers (including widows) for the landmark Belgaum Congress Session of 1924. As public fury over the Rowlatt Act, 1919, swept the country, a young Umabai, only 27 years old, knew this was her calling. Gandhi had appealed to Indians to fight this unjust act. On 13th April 1919, the country was shocked when around 400 unarmed protestors were killed and more than 1200 people were injured in Amritsar's Jallianwala Bagh for peacefully resisting the law. Umabai was one of the many who took to the streets this time to voice their outrage at the tragedy.

Umabai Kundapur's struggle for India's independence was about much more than defeating colonial power. Without social reforms such as education, cheap healthcare, and women empowerment, independence meant little to her. Umabai Kundapur was born in Mangalore in 1892 to Golikeri Krishna Rao and Jungabai. She was married to Sanjiv Rao

Kundapur, from a well-to-do family. Anandarao Kundapur, her father-in-law, was a progressive thinker who believed in raising women's status. Umabai pursued her schooling under his guidance and encouragement and passed her matriculation examination. After completing her education, Umabai helped her father-in-law in educating women through Gaundevi Mahila Samaj in Mumbai.

The early twentieth century was an era of awakening, and cities like Mumbai saw growing protests against British atrocities. Mahatma Gandhi and the Congress called for a nationwide strike in protest of the Rowlatt Act, which granted the British government full authority to arrest and detain anyone.

Since the early 1900s, Mumbai has been a hub of political activity. The Sardar Griha served as the protest headquarters, and Lokmanya Tilak spearheaded the protests. Tilak was a major part of the liberation struggle and this is evidenced by the fact that his funeral gathered about two hundred thousand people to Mumbai's Chowpatty beach a year later in 1920, including Umabai and her husband. Umabai decided to volunteer for Congress at this point. Her in-laws, as well as her husband, Sanjiv Rao, supported Umabai in her political endeavors. Unfortunately, tragedy struck in 1923, when Sanjiv Rao died of tuberculosis.

A distraught Umabai relocated to Hubli in Karnataka, where she was encouraged to manage the family-owned Karnataka Press and the girls' school 'Tilak Kanya Shala'. Soon after, she recognised the dearth of female participants in marches and demonstrations, she began mobilising women via door-to-door campaigns. She co-founded the

'Bhagini Samaj', a women's NGO, along with another nationalist, Krishnabai Panajkar. She chose to educate other women after her education and literacy experiences through this effort.

Meanwhile, Umabai remained a member of the Congress and travelled across Karnataka, encouraging women to quit their confinement and join the freedom movement. Her efforts to inspire children and their parents to join the independence struggle were noted by freedom fighter Dr N S Hardikar, the founder of the Hindustani Seva Dal in 1924. He appointed her as head of the Hindustani Seva Dal's women's branch.

Umabai coached several young people in spinning, weaving, camping, and drills as part of her employment. As the Dal grew in membership and activity, national leaders, including Gandhi, were impressed by her work and commitment. The idea behind enrolling women in public camps for physical training and marches was previously unheard of, like a revolutionary idea at the time. However, Umabai was successful in rallying women and gaining followers with novel approaches such as plays and theatre.

The All-India Congress Committee conference was hosted in Belgaum in 1924. It made history not just by holding the session, but also by providing the momentum and direction for our struggle for independence. Mahatma Gandhi presided over it. Umabai's efforts to bring together women from various sections of Karnataka paid off when she was one of the session's organisers. She made a mark with her sheer organisational skills. The session gathered a large number of women from the impoverished Dharwad district, including widows with tonsured heads who had been relegated to household labour then. Umabai overcame a near-impossible mission when she mobilised over 150 women volunteers for the landmark Congress Session. She influenced women like Kamaladevi Chattopadhyay, who went on to play a very important role in the Salt Satyagraha in Mumbai years later and was fascinated by the soft-mannered and iron-willed Umabai.

Umabai was a key figure in the Salt Satyagraha in the early 1930s, for which she was arrested in 1932 and imprisoned for four months in Yerwada jail. Meanwhile, the British seized her press, sealed her School, and declared her NGO to be illegal. Undaunted by the events, Umabai decided to continue battling. Her modest home was transformed into a haven for female freedom warriors. She not only accommodated them but also provided them with funds for their return journey.

On the day she was released from prison she got to know that her father-in-law had died a week earlier. This was an irreparable painful setback for Umabai. Since her childhood, she was fortunate that her father-in-law had been her most ardent supporter, encouraging academic pursuits, social work, and even political activism. He was unlike the conservative people of those times.

Umabai decided that rededicating herself to social service was the most fitting way to honour her late father-in-law. People (including women) were imprisoned during the Salt Satyagraha in 1930, and when they were released, they were frequently homeless because their families had abandoned them due to the fear of the British. They took refuge at her home. Again, following the devastating Bihar earthquake of 1934, Umabai and her volunteers came forward to work day and night in refugee camps. She met national leaders such as Rajendra Prasad and Acharya Kriplani during this period.

She fearlessly offered her home to stay to a number of freedom fighters, escaping arrest. Mahatma Gandhi appointed her head of the Kasturba Trust's Karnataka branch in 1946. The Kasturba Trust was founded to develop villages through the training of volunteers known as Grama Sevikas to educate, conduct health programs, and oversee child welfare. Following independence, she was offered major political positions, but she did not accept all of them, saying that her vow was service to the nation. She even refused to accept the prizes, accolades, and pensions accorded to freedom fighters. She continued to labour and eventually retired to a small home in Hubli, where she died in 1992 at the age of one hundred.

To become a legend, one needs to showcase their fearlessness and passion. Umabai was s fearless leader who sheltered freedom fighters from the British, who did not succumb to the cruel blows of destiny, who resolved to fight for the right.

RAMADEVI CHOUDHURY
(1899–1985)

16

RAMA DEVI CHOUDHURY

Stern as Steel

Rama Devi Choudhury, one of the most admired names in Odia history, was among the women who dedicated themselves to the freedom struggle. Despite being from a wealthy family, she gave up all her luxury and comfort for the sake of independence. Among her, numerous contributions were her participation in Salt Satyagraha. Choudhury devoted a substantial portion of her life to Harijan welfare, community improvement, and the elimination of untouchability. Her participation in social movements before India's independence was mostly focused on opposition to the colonial rule, she addressed economic injustices through these land reform initiatives. She believed in the all-encompassing meaning of swaraj, or self-rule, based on Gandhian ideas.

Among the women freedom fighters in Orissa Rama Devi occupied a unique and unparallel position and played an outstanding role in the freedom movement of the state. Though she belonged to a very rich zamindar family she abandoned all her luxury and comfort for the cause of independence. After joining National Congress, she used to wear only khadi leaving apart the luxury sarees, clothes and ornaments. In her first meeting with Mahatma Gandhi at Cuttack in the year 1921 in Orissa, she offered a hand-woven cotton bundle to Mahatma Gandhi when he could visualize the extraordinary spirit of the woman for nationalism and patriotism.

Choudhury was born in Cuttack on December 3, 1899, to Gopal Ballav Das and Basant Kumari Devi. Her aristocratic family background, as well as her uncle Utkal Gaurab Madhusudan Das—a distinguished lawyer, social reformer, and public personality—inspired her career as a freedom fighter and social reformer. Born during the Swadeshi Movement, her life was heavily influenced by the teachings of the philosopher Sri Aurobindo, as well as activists and leaders like JP Narayan and Vinoba Bhave. Rama Devi married Gopabandhu Choudhury, a Deputy Collector. In 1921, he resigned from his government post to join the independence struggle with his wife. They joined the Indian National Congress the same year. Rama Devi was deeply moved to join the freedom cause after witnessing a conference of forty women at Cuttack's Binod Vihari Temple. Gandhi spoke at the assembly, and his non-violent techniques were particularly popular among Odia women. Furthermore, this conference was the first time Odia women attended a public gathering, as well as the first time a leader of great political standing, such as Gandhi, addressed Odia women in a separate assembly.

In the year 1930, she was an active participant in the Salt Satyagraha in several parts of Orissa. Due to the British Government's harsh prohibition in Cuttack to continue the protest, she travelled to Inchudi and Srijang in Balasore District with Kiranbala Sen and Malati Devi, where thousands of women joined them in the Salt Satyagraha movement, also known as Inchudi Satyagraha. Rama Devi and Pranakrushna Padhiari were arrested at Swaraj Ashram in November of the same year, following the signing of the Gandhi-Irwin Pact, they were released from jail.

She got active in Harijan welfare, community improvement, and the elimination of untouchability. She founded the Asprushyata Nibarna Samiti, later renamed the Harijan Sewa Sangh, for this purpose. Almost fifteen women joined in the work against untouchability. They focused their efforts on mothers, whom they viewed as the most potent agents of change. In 1934, numerous women, including Rama Devi, joined Gandhi's Harijan Padyatra in Puri. Gandhi regarded Rama Devi with special reverence and admiration.

"I have marvelled at how Smt. Rama Devi and her girls have discharged themselves during the tour, which was undoubtedly pleasant and easy, had undoubtedly its trials. But these ladies have not known what fatigue is? Nowhere have I seen anything quite like what Smt. Rama Devi and her little band have been found to do so gracefully and so naturally. They have never needed or claimed any special privilege".

Rama Devi was also deeply engaged in the Odisha trips of Jawaharlal Nehru, Kasturba Gandhi, Maulana Azad, Rajendra Prasad, and Sardar Patel. Following Gandhi's dictum, Rama Devi and her husband established the Sewaghar ashram in Bari. Through numerous programmes, the ashram prioritised the social and economic improvement of rural folks. In the Ashram volunteers were engaged in a variety of activities, including promoting Khadi, anti-untouchability movements, cleaning the toilets, self-services, education, adult education, awakening women to various issues, basic education and the spread of the national language, drug de-addiction programmes for adults, increasing self-employment of common men through dairy development, beekeeping, leather works and delivering health services to the patients, etc.

In 1942, the Quit India Movement led to the imprisonment of Rama Devi and twenty-three other women in Odisha. Additionally, the British deemed the ashram unlawful. After her liberation in 1944, Rama Devi devoted herself to advancing Khadi, training and educating women, and basic education. In the year following Kasturba's death, Mahatma Gandhi appointed Rama Devi to oversee the Kolkata branch of the Kasturba Trust.

After India's independence, Rama Devi devoted her career as a reformer to Vinoba Bhave's Bhoodan and Gramdan campaigns, which pursued reform through land grants from rich landowners to landless individuals. During the 1950s and 1960s, she joined Bhave's Sarvodaya movement, which advocated people's self-organisation. Rama Devi and her husband walked 4,000 kilometres to spread the message of donating land to the landless and the impoverished in different corners of Orissa like Ganjam, Gunpur, Bisam Cuttack, Bhawanipatna, Sambalpur, Balangir, Brahmagiri, etc.

In Ramchandrapur, she also founded the Utkal Khadi Mandal, a teachers' training centre. She founded a Tribal

Welfare Centre in 1950 at Dumburugeda. She assisted families affected by the famine in 1951 and participated in relief efforts in Kolatpur. Despite her exhaustion and stress, Rama Devi enthusiastically assisted the soldiers afflicted by the 1962 Indo-Chinese War. Rama Devi demonstrated her opposition to the government during the Emergency by publishing her newspaper, The Gram Sevak Press.

In her later years, Rama Devi also created the Shishu Vihar primary school and the Cuttack cancer hospital. She was nominated for the Padma Shri in recognition of her ceaseless efforts toward community welfare and her extraordinary contributions to the independence fight. However, Rama Devi declined the honour as well as countless other similar awards from various organisations. She believed that her efforts to end infringement and injustice were ongoing, and didn't need honouring.

She died on July 22, 1985. However, her greatness endures through the lives she reformed. In her honour, the Rama Devi Women's University in Bhubaneshwar bears her name. It is the first university for women in eastern India, having been founded as such in 2015. There is a museum dedicated to her on the campus of the university. Along with her contemporaries, Rama Devi guided India on an equitable path to independence, as well as to a liberal, caste-free future.

When cause is above than the comfort and wealth, it hardly seems sufficient tribute to someone who could have lived a life of amenity but chose an incredibly difficult, but free life in its place. The lessons from the Rama Devi Choudhury, still pertinent to this day.

KAMALA NEHRU
(1899–1936)

17

KAMALA NEHRU

A 'shy' wife to strong fighter

While we all know Jawaharlal Nehru's contribution to the Indian independence movement, not many are aware of his wife, Kamala Nehru's role in the Indian freedom movement. She was courageous, daring and determined. Kamala Nehru, the wife of Jawaharlal Nehru and an ardent devotee of Mahatma Gandhi, was among the first freedom warriors to advocate for women's empowerment. After marrying into the Nehru family, despite having been raised in a traditional home, Kamala Nehru quickly emerged as a formidable force in the national movement for freedom.

Kamala Nehru was born in 1899 into a Delhi-based family of Kashmiri Brahmins. The eldest daughter of Rajpati and Jawaharmal Mull Atal-Kaul, Kamala Nehru was also the first lady of India. She had three brothers and a sister. Kamala Nehru was a reserved young lady, largely because she was born into a modest family and raised in a conservative environment. Though the westernized Nehru family was a culture shock for Kamala Nehru, her feelings of alienation were quickly overcome by a desire to work for India's independence. On February 7, 1916, Kamala Nehru married Jawaharlal Nehru in a traditional Hindu wedding ceremony. At the time, Jawaharlal Nehru led the Indian National Congress. As a result of being homeschooled and her Indian heritage, Kamala Nehru was unable to speak English at the time of her marriage, unlike her husband. On November 19, 1917, following her marriage to Jawaharlal Nehru in 1916, Kamala Nehru gave birth to her first child, Indira Priyadarshini.

Kamala Nehru had a strong relationship with Kasturba Gandhi, Mahatma Gandhi's wife, and also spent some time at her ashram. In this ashram, Kamala Nehru forged a close bond with the Bihar freedom fighter Prabhavati Devi.

The Non-Cooperation Movement of 1921 marked Kamala Nehru's entrance into the country's freedom struggle. Once considered a timid personality, Kamala Nehru overcame all preconceptions and joined her husband in the Indian Independence struggle. She, along with other female pioneers, led a big protest against the shops in Allahabad that sold foreign cloth and alcohol. It is said that Kamala Nehru pushed Jawaharlal Nehru to alter his lifestyle after she was profoundly influenced by Mahatma Gandhi's ideals.

Following in the footsteps of Mahatma Gandhi, she joined the civil disobedience movement to fight the British salt monopoly. During the salt Satyagraha, she was thought to have been one of the first leaders to sell illegal salt. When Jawaharlal Nehru was arrested by the police, allegedly to prohibit him from delivering a "seditious" speech, his wife Kamala Nehru took over. She assumed the stage and delivered his speech in his place. The British placed Kamala

Nehru in jail out of concern for her growing popularity. She boldly braved the lathi attacks and continued to organise demonstrations with other female freedom activists in different locations.

Even while her husband, Jawaharlal Nehru, was imprisoned for months, Kamala Nehru continued to fight for freedom and established a clinic for wounded freedom fighters at the Swaraj Bhawan, the Nehru residence.

She encouraged women from conservative backgrounds to participate in picketing fabric shops and selling Khaddar on the streets in defiance of the government ban, despite facing objections. Together with other women volunteers, like Durgabai and Kamaladevi Chattopadhyay, she also organised anti-tax campaigns.

Krishna Nehru Hutheesing (Indian writer, the youngest sister of Jawaharlal Nehru and Vijaya Lakshmi Pandit) wrote, "Kamala was a source of comfort to Jawahar and never let him know how ill she had become and how much she needed him. She was always optimistic, and whenever she was near Jawahar, she soothed him and infused new courage into him after a moment of anxiety and disillusionment."

She had been ill with tuberculosis for a very long time, which led to her premature demise on February 28, 1936, at the age of 37. Her contributions to the Indian Freedom Movement are acknowledged and valued even today. In recognition of Kamala Nehru's contributions to the well-being of independent India, a great number of institutions and places of public significance in Delhi were built in her honour—the University of Delhi's Kamala Nehru College, the Kamala Nehru Hospital, and the Kamala Nehru Park were constructed. She left behind countless tales of fortitude and tenacity that made her an amazing woman.

Learning to break stereotypes and stepping out of their comfort zone are some attributes learnt from Kamala Nehru. It also requires inner strength, willpower and passion to transform self and others, and Kamala Nehru did it with composure while overcoming her fears.

MANIBEN PATEL
(1903–1990)

18

MANIBEN PATEL

A Vintage Heart with a Beautiful Mind

Sardar Patel's daughter, Maniben, was a diligent aide with a self-effacing attitude. Maniben actively participated in the Non-Cooperation Movement and the Salt Satyagraha. She was imprisoned multiple times for her participation in the freedom movement. Her father instilled in her a strong sense of patriotism. She contested the first Lok Sabha election in 1952 from the Kheda (South) constituency and won by a margin of 59,298 votes.

Maniben was born on April 3, 1903, in Karamsad, Gujarat, and lost her mother when she was only six years old. Vithalbhai Patel, her uncle, took responsibility for her upbringing. She attended The Queen Mary High School in Bombay and graduated from Vidyapith, which Mahatma Gandhi founded in Gujarat. She followed Mahatma Gandhi's ideas and began working frequently at his ashram in Ahmedabad in 1918. Maniban went on to help her father after graduating in 1925.

In 1923-24, the British government imposed high taxes on the common people and began taking their cattle, land,

and property to collect the money. Maniben inspired women to join a movement led by Gandhi and Sardar Patel to support the No-Tax Movement to fight this discrimination. The British authorities collected exorbitant taxes on the peasants of Bardoli in 1928, and they were subjected to similar abuse as those of Borsad. Mahatma Gandhi appointed Sardar Vallabhbhai Patel to spearhead the Satyagraha. Women were first hesitant to join the campaign. Maniban, Mithuben, and Bhaktiba Desai inspired women who eventually surpassed men in the campaign. As part of the protest, they stayed in huts built on government-seized land.

Maniben soon took charge of Patel's appointments and day-to-day activities. She escorted her father everywhere and was well-versed in his opinions on every matter that he became involved in. Maniben kept a diary of her father's, day-to-day activities.

A Satyagraha was planned in 1938 to protest the Diwan of Rajkot State's harsh rule. Despite her terrible health, Kasturba Gandhi was eager to join the Satyagraha, and Maniban joined her. The administration issued an order segregating the women. She went on hunger strike in protest of the order, and officials eventually permitted her to reconnect with Kasturba Gandhi.

Maniben played an active role in the Non-Cooperation Movement and the Salt Satyagraha. She was imprisoned multiple times for her participation in the freedom movement. She was imprisoned in Pune's Yerwada Jail from 1942 until 1945 as part of the Quit India movement. She began her day in jail with prayers, as a severe disciplinarian like her father and a staunch Gandhian, followed by spinning, reading, strolling, washing, repairing, and caring for the ailing inmates.

Maniben Patel was devoted to her father until he died in 1950. She later served as a member of the Indian National Congress led by Nehru in the first Lok Sabha (1952–57) from the South Kheda seat and the second Lok Sabha (1957–62) from the Anand constituency. She was also Secretary of the Gujarat State Congress (1953–56) and Vice President (1957–64). She was elected to the Rajya Sabha in 1964 and served until 1970. Maniben Patel was jailed in 1976 for her protest demanding the release of leaders, the end of the Emergency and the abolition of press censorship. She also led another Dandi March in 1976 to raise awareness and courage among the people. She spent the rest of her life in Mumbai working for several philanthropic organisations including the Sardar Patel Memorial Trust. She went on to write a book about her father's life in the years following Indian independence, which included an account of the freedom movement.

She was involved with several social and educational organisations, including Gujarat Vidyapith, Vallabh Vidya Nagar, Bardoli Swaraj Ashram, and Navjivan Trust. Her illustrious career, after almost three decades of being an MP, came to an end in 1990. In 2011, the Sardar Vallabhbhai Patel Memorial Trust collaborated with Navajivan Publications to publish her Gujarati diary.

A galaxy of extraordinary compassion is enough to mark your love and sacrifice for the Motherland. Someone like the Exemplary Maniben Patel, daughter of the "Iron Man of India" Sardar Vallabhai Patel is hard to find in today's age. A woman who knew how to maintain her dignity and self-respect in all the circumstances.

KAMALADEVI CHATTOPADHYAY
(1903–1988)

19

KAMALADEVI CHATTOPADHYAY

When words are spoken
straight from the core

Kamaladevi Chattopadhyay is one of the most admirable feminist liberation fighters, social reformers, art aficionados, social activists, performers, youth leaders and forward-thinkers who reshaped our nation's image in all these fields. She was the driving force behind the resurgence of Indian handicrafts, handlooms and theatre in independent India, and for pioneering cooperation to improve the socioeconomic status of Indian women. Asked by a journalist about her views on women's empowerment, replied, "Let men learn to be equal to women first". For breaking Salt laws, she was sent to jail by the British. Chattopadhyay authored over twenty books, many of which were inspired by her personal travel experiences. After Partition, she worked relentlessly to establish the city of Faridabad as a resettlement centre for immigrants from the Northwest Frontier province. Further, she founded the Indian National Theatre (INT) in 1944, which evolved into what is currently known as the National School of Drama.

Kamaladevi was born on April 3, 1903, into a liberal Saraswat Brahmin family in Mangalore, Kamaladevi was the fourth and youngest daughter of Ananthaya Dhareshwar (a district collector in South Kanara district of the then-Madras Presidency) and his wife Girijabai. Her primary schooling was done at a neighbourhood convent school. She developed an appreciation for traditional art forms as a result of growing up in a place with a rich cultural legacy, particularly for the music and dance form Yakshagana. Tragic events occurred early in her life when her elder sister Saguna, whom she regarded as a role model, died in her teens, shortly after her early marriage, and her father died when she was only seven years old. To make matters worse for her mother Girijabai, he died without leaving a will for his large estate and the ownership of all his properties was transferred to his son from his first marriage. This left Kamaladevi's mother in a difficult circumstance, as they were expected to survive on a monthly allowance. Girijabai declined the allowance stubbornly and chose to raise her girls on her own. Thus, Kamaladevi was raised in the home of her maternal uncle, a prominent social reformer. Political intellectuals and popular personalities such as Gopalkrishna Gokhale, Sir Tej Bahadur Sapru, Mahadev Govind Ranade, Srinivasa Sastri, Annie Besant, and Pandita Ramabai frequently paid him visits. Kamaladevi was a little child and would often meet these famous people and her experiences with them sowed the seeds of political awareness in her mind. Her educated mother and industrious grandmother, on the other hand, made the strongest effect on her psyche. They were the ones who instilled in her an independent spirit and a lifetime love of literature.

She got married in 1917 to Krishna Rao, but was widowed two years later. Her father-in-law was liberal in outlook and pushed her to complete her schooling. She took his counsel to heart and committed the next few years to study. She continued her education at Queen Mary's College in Madras (now Chennai). There, she met Harindranath Chattopadhyay, Sarojini (Chattopadhyay) Naidu's brother. Harindranath, a vivacious poet, writer and actor, shared common interests with Kamaladevi, including a fondness for the arts and a love of music and theatre. The pair quickly felt a bond and married, despite considerable opposition from the orthodox community (as Kamaladevi was a widow). The following year, Rama, their only child, was born. Kamaladevi toured India with her husband, explored folk theatre and regional drama, and even appeared in silent films. She accompanied Harindranath to London and joined a diploma study in Sociology at Bedford College, University of London. Their marriage, however, ended due to incompatibility concerns, and they later divorced cordially, with their divorce reportedly marking India's first legal separation allowed by the courts.

Kamaladevi learned of Gandhi's Non-Cooperation Movement in 1923 while still in London. and she returned to India shortly thereafter to join the Seva Dal (a Gandhian organisation dedicated to the social improvement of the destitute) and to register with the Indian National Congress. Soon after, she was appointed head of the Dal's women's division, where she became active in enrolling, training, and mobilising girls and women of all ages across India to work as volunteer sevikas'.

Three years later, Kamaladevi earned the unique distinction of becoming the first woman in India to run for political office. She was close to Margaret Cousins, the Irish theosophist feminist and founder of the All India Women's Conferences. She later became the AIWC's Standing Committee's Organising Secretary. Margaret also motivated her to run for the Madras Provincial Legislative Assembly. Though she had only a few days to campaign, she ran for a seat in the Madras Legislative Assembly and was defeated by a slim margin of 55 votes. At this pivotal point in history, Kamaladevi possessed the expertise and enthusiasm necessary to assume the role of a nationalist leader.

In the years that followed, AIWC developed into a reputable national organisation, with branches and volunteer programmes located throughout the country, and an unwavering commitment to legislative reform. During her term, she travelled widely throughout Europe, where she was inspired to begin numerous social reform and community welfare projects, and establish educational institutions managed by and for women. Another remarkable example of this is the establishment of Lady Irwin College for Home Sciences in New Delhi, a first-of-its-kind college for women in its day. Kamaladevi began writing about women's rights in India in 1929; one of her most recent works, Indian Women's Battle for Freedom, was released in 1982. Over a five-decade span, Kamaladevi established a distinct viewpoint in dozens of writings and speeches, one that was conscious of the vulnerabilities faced by Indian women that were both unique to them and universal to women worldwide.

Despite her enthusiastic support for the Salt Satyagraha, Kamaladevi disagreed with Gandhi's decision to ban women from the march. She pushed Gandhi to allow women equal access to the march's front lines. Later, she was a member of the seven-person lead team designated by Mahatma Gandhi in the famous Salt Satyagraha (1930) to produce salt on the Bombay beachfront; Avantikabai Gokhale was the team's sole other female volunteers. Later, in an unexpected move, Kamaladevi approached a local High Court and asked a magistrate if he would be interested in purchasing the "Freedom Salt" she had just produced, also attempted to sell packets of 'freedom' salt at the Bombay Stock Exchange and was sentenced to prison for violating the salt rules. On 26 January 1930, she made national headlines when, after a struggle over the Indian Flag, she clutched it determinedly to preserve it from British forces. Kamaladevi's prominence as a participant and her direct boldness aided in recruiting hundreds more women volunteers. She became the Congress Socialist Party's president in 1936, with Ram Manohar Lohia and Jayaprakash Narayan.

From the late 1920s to the 1940s and beyond, Kamaladevi served as an ambassador and advocate for Indian women and political independence and larger transnational causes such as the emancipation of coloured people worldwide from colonial rule and the promotion of political and economic equity between nations. Kamaladevi Chattopadhyay travelled to Denmark in 1939 to attend a conference on women's rights and was in England at the outbreak of World War II. She immediately began travelling to several countries to raise awareness about India's plight and to rally support for India's independence struggle. She travelled to America and talked with many political activists, largely black, sharing India's non-violent commitment to gain freedom. She made friends with American feminists, engaged in dialogue with African-American civil rights activists and contributed to various American journals. After learning of her activities, the British barred her from returning to India. Kamaladevi continued her voyage unafraid, stopping at South Africa, China, Japan, and Vietnam.

Kamaladevi was a brilliant author, and her writings attest to the breadth of her intellectual and political interests, as well as a global view that opposed both limited nationalism and shallow cosmopolitanism. She visited Nanjing and Chongqing and talked with resistance leaders during the country's Japanese occupation—this culminated in the publication of War-Torn China (1944). In keeping with her inquisitive attitude, she travelled to Japan and wrote Japan: Its Weakness and Strength (1944) highlighting how the Japanese stained their hands with the most aggressive strains of materialism and imperialism. She is also one of those in India who wrote extensively about the United States. She redirects the perspective in Uncle Sam's Empire (1944) and America: The Land of Superlatives (1946).

Kamaladevi also rehabilitated thousands of post-partition refugees who had flocked to Delhi, primarily from West Punjab, in quest of shelter and work and were living in improvised tents throughout the city. With the approaching Delhi winter certain to exacerbate the situation, Kamaladevi decided to form the Indian Cooperative Union. Despite enormous challenges, ICU was able to construct an industrial township in the suburbs of Delhi called Faridabad. Following independence, she devoted her efforts to revitalizing India's handicrafts and textiles. She founded several of the nation's first institutions dedicated to collecting, preserving, and promoting Indian dance, theatre, art, puppetry, music, and handicrafts, like the National School of Drama, Bharatiya Natya Sangha, Sangeet Natak Academy, Central Cottage Industries Emporium, World Craft Council, Craft Council of India, and Delhi Craft Council.

However, feminism remained a cause important to her heart. Throughout her long involvement in the liberation struggle, she never steered away from confronting her colleagues on women's rights, even if they were stalwarts like Motilal Nehru, C. Rajagopalachari, and Mahatma Gandhi. Kamaladevi was unquestionably the epitome of female empowerment. Her remarriage following widowhood and formal divorce from her second marriage was symbolic of her reclaiming her independence.

Kamladevi acted in films during a time when the film business was not regarded as a respectable place for women. She worked in Tansen, starring K.L. Saigal and Khursheed, Shankar Parvati (1943), and Dhanna Bhagat (1945), having begun her cinema career in 1931 with the Kannada film industry's first silent film, Mricchakatika (Vasantsena). The government of India honoured her with the Padma Bhushan in 1955 and the Padma Vibhushan in 1987. In 1966, she received the Ramon Magsaysay Award for Community Leadership. In 1974, she was conferred with the titles of Ratna Sadsya by the Sangeet Natak Akademi and Desikottama by Shantiniketan, the highest awards of both organisations. She was also recognised for her contributions to the promotion of handicrafts by UNESCO, UNIMA (Union Internationale de la Marionnette), the International Puppetry Organisation and the World Crafts Council.

Kamaladevi died on 29 October 1988 and left an indelible imprint of her intellect and deep insights into everything she touched. Her contribution to so many diversified areas of human life is incomparable.

You need to have a serpent's tongue if you are not heard because of your gender and have courage to speak out what you believe in. Kamala Devi is a great example to highlight this and put light on the importance of the choices made at every stage being decisive, when there is a challenge to be overcome.

VIJAYA LAKSHMI PANDIT
(1900–1990)

20

VIJAYA LAKSHMI PANDIT

A brave woman who set a higher goal

Vijaya Lakshmi Pandit was a participant in the country's freedom movement and worked as a diplomat before embarking on a political career. She actively participated in the freedom struggle, as a result of which she was incarcerated thrice, that included criticizing the Emergency. She was Prime Minister Jawaharlal Nehru's younger sister and was a freedom fighter who held significant national and international positions. She had a certain grace, courage, brilliance and determination that made her admirably fitting for high duty. Mohandas Gandhi had a considerable influence on her. Vijaya Lakshmi Pandit was elected to the Legislative Assembly of the United Provinces in 1937 but resigned in 1939 in protest over the British involvement in World War II. She was the first woman to be elected President of the United Nations General Assembly in 1953, when she presided over the eighth session, besides holding several significant diplomatic positions.

Vijaya Lakshmi Pandit was born as Swarup Kumari Nehru to the Nehru family on 18 August 1900 in Allahabad. Motilal

Nehru, her father, was a renowned lawyer, and political independence fighter. She had her education exclusively at home and in Switzerland, with instructors overseen by her English governess. She was stunning, brilliant, witty, at ease with people of all social classes and, sometimes, a little sharp-tongued.

As was customary in her family, she became an active member of the Indian nationalist movement too, when Motilal Nehru backed Mohandas Gandhi, the head of the Indian National Congress, in his fight against British rule. Vijaya married Ranjit S. Pandit in 1921, at that time she changed her name to Vijaya Lakshmi Pandit. The couple had three daughters. She and her husband were both involved in the independence struggle and were imprisoned at various points. He died in 1944, following a prison sentence.

Pandit's long career in politics and foreign diplomacy began in 1934 when she was elected to the Allahabad Municipal Board. She was elected to the Assembly of the United Provinces in 1936 and appointed minister of local self-government and public health in 1937, making her the first Indian woman to serve in a cabinet position. As with all other Congress party officials, she resigned in 1939 in protest of the British government's declaration of India's participation in World War II. Between 1941 and 1943, Pandit served as president of the All-India Women's Conference, encouraging public opinion in support of gender equality and women's welfare. She and her husband, along with other Congress leaders, were imprisoned by the British in 1942, during the height of the Quit India campaign.

Pandit travelled to the United States, after being driven to re-focus her life following her husband's death, in the year 1945. Her trip to the states was primarily a lecture tour and a visit to see her daughters, Nayantara and Chandralekha, who were students at Wellesley College. She talked extensively about India, becoming acquainted with prominent Americans, including Eleanor Roosevelt. She restarted her post as minister of local self-government and public health in the United Provinces upon her return to India in January 1946.

She embarked on her first official diplomatic journey as the leader of the Indian delegation to the United Nations General Assembly in the fall of 1946, and again in 1952-53. Shortly after India's independence in 1947, she joined the foreign service and was appointed India's first ambassador to the Soviet Union (1947–1949) and then to the United States (1949–1951). Additionally, she served as the leader of India's delegations to the United Nations General Assembly in 1947, 1948, 1952, 1953, and 1963.

Pandit became the first female president of the United Nations General Assembly in 1953. She presided over the General Assembly's eighth session. She was the Indian high commissioner (ambassador) to the United Kingdom for nearly seven years, commencing in December 1954, throughout a tumultuous period in British-Indian ties during the Suez and Hungarian crisis in 1956.

From 1962 to 1964, she served as governor of Maharashtra and as a member of Parliament from 1964 to 1969. She was shocked by Jawaharlal Nehru's death on May 27, 1964. She was elected to the Lok Sabha in November following a by-election in Uttar Pradesh's Philpur district. She was re-elected in 1967's fourth general election, but resigned the following year due to "personal reasons."

Although Vijaya stepped away from politics in the late 1960s, her dedication to freedom drew her back into the Indian political scene in the 1970s. Furious at Indira Gandhi's state-of-emergency suspension of democratic processes from 1975 to 1977, she campaigned against her niece, Indira Gandhi. Her efforts resulted in Gandhi's election defeat. Due to increasing strains, Pandit joined the opposition party and was elected to Parliament in 1978, defeating Indira Gandhi. In 1978, she also served as India's delegate to the United Nations Commission on Human Rights.

'So I Became a Minister' (1939) and 'Prison Days' are among Pandit's writings (1946). Vijaya Lakshmi Pandit died on December 1, 1990, in Dehradun, India. Newspapers lauded her as one of the century's most outstanding women. Vijaya Lakshmi Pandit, a diplomat and politician of distinction, achieved remarkable heights and left her imprint. Throughout her career, she fought against gender segregation and emphasised the importance of men and women working together to establish a nation. She rose to prominence as India's ambassador to the world.

It requires courage to be an empathetic, elegant taugh and compassionate woman when you want to change people's mindsets and Vijaya Lakshmi Pandit life showcases exactly what was needed to establish growth while maintaining the objectivity and fairness in diverse situations.

USHA MEHTA
(1920–2000)

21

USHA MEHTA

Your voice matters—42.34 metres from somehwere in India

Usha Mehta, known as a young revolutionary, participated in a protest march against the British when she was only eight years old. Even though her father, a judge under the British Raj, did everything he could to prevent her from joining the liberation struggle, she defied all restrictions to fight for her homeland. She joined the Quit India Movement after dropping her studies and devoted herself to the campaign. During the few months of the Quit India Movement, Usha Mehta broadcasted the underground radio station which broadcast stirring bulletins from secret locations to counter the propaganda and the voice of the British administration. Usha Mehta, a college student was only 22 years old when she went "underground" to operate a secret radio station during India's struggle for independence from British rule. Ultimately, the Raj discovered the radio's location and arrested its initiators. Usha Mehta herself spent several years in jail. "When the press is gagged and all the news banned, a transmitter certainly helps a good deal in spreading the message of rebellion in the remotest corners of the country." Her time behind the radio microphone was brief, but its influence was significant.

Born on 25 March 1920 in the village of Saras in the Gujarat region of Surat, not only was Gujarat Gandhi's home state, but it was also the site of his iconic salt march in 1930. When she was just five years old, Usha first saw Gandhi while on a visit to his ashram at Ahmedabad. Soon, Gandhi organised a camp near her village in which Usha participated, attending sessions and doing a little spinning. Usha's father was a British Raj-era judge. Consequently, he did not push her to join the independence struggle. However, this restriction was lifted in 1930 when her father retired. Usha's family relocated to Bombay in 1932, when she was 12 years old, allowing her to participate more actively in the liberation fight. She and other children disseminated secret pamphlets and periodicals, visited relatives in jails, and delivered messages to these inmates. She was a teenager when she responded to Gandhi's call to defy the salt tax; broke the law and did something for the nation even as a young child.

Usha received her early education at Kheda and Bharuch before attending Chandaramji High School in Bombay. She was of average intelligence. In 1935, her matriculation exams placed her among the top twenty-five pupils in her class. She completed her education at Wilson College, Bombay, graduating with a degree in philosophy in 1939 with the highest grade. She began studying law as well, but in 1942 she abandoned her studies to join the Quit India Movement, and participated in numerous civil disobedience initiatives, including picketing, rallies, and spinning cotton to resist British imports. Then, beginning at the age of 22, she engaged in the liberation movement full-time.

Usha grew up under Gandhi's influence and became one of his followers. She made an early resolution to be chaste for life and adopted a minimalist, wearing simply Khadi clothing and avoiding all forms of luxuries. Gandhi had declared that the Quit India Movement would begin with a gathering at Gowalia Tank grounds in Mumbai on August 9, 1942. Before that day, nearly all leaders, including Gandhi, were arrested. On the scheduled date, however, a large number of Indians gathered at Gowalia Tank Ground. It fell to a group of subordinate leaders and workers to deliver the address and raise the national Flag.

On 14 August 1942, Usha and many of her close allies founded Secret Congress Radio, a covert radio station. It aired on August 27. The first words broadcast in her voice were: "This is the Congress Radio calling on (a wavelength of) 42.34 meters from somewhere in India". This is how the broadcast would always begin (Mehta later revealed that they were all recorded in Bombay). She was able to launch the station with the assistance of two other activists, Chandrakant Babubhai Jhaveri and Vithaldas K Jhaveri, as well as Nanka Motwane, whose family controlled Chicago Radio, a telephone firm. Additionally, amateur radio operator Nariman Printer also assisted. Initially, they were broadcasting twice every day in both Hindi and English. However, they curtailed it to a single evening occurrence. The Radio's ability to ignite the torch of independence in its listeners' hearts and encourage them during those dismal and tough times was astonishing.

Moreover, it conveyed to the youth the great worth of aspirations and dreams, as well as the significance of voluntary and laborious efforts in making these highly improbable dreams a reality.

Her associates consisted of Vithalbhai Jhaveri, Chandrakant Jhaveri, Babubhai Thakkar, and Nanka Motwani, the proprietor of Chicago Radio, who supplied equipment and technicians. In addition to Dr Ram Manohar Lohia, Achyutrao Patwardhan, and Purushottam Trikamdas, other leaders, aided the Secret Congress Radio. However, they continued to travel to throw the authorities off their scent; they changed locations between six and seven times throughout the three months they broadcasted. The channel broadcasted a wide variety of stories, including merchants refusing to export rice and the arrests of leaders and people. Even though it was located in Bombay, Congress Radio's influence extended well beyond the city's borders; it encouraged freedom-lovers throughout the nation.

The authorities raided radio stores in Bombay in November 1942, including one owned by Chicago Radio. They arrested Nariman Printer, who was suspected to have provided information on the location of the radio station. On the evening of November 12, after Mehta relayed the program and played the 'Vande-Mataram' record to end the show, the broadcast was interrupted. The police stormed Babubhai Khakkhar's office while she was present. She was detained along with four others, and the inquiry lasted months. Three of the five were convicted, including Mehta. She was given a four-year prison sentence and freed in April 1946. Six months were devoted to her interrogation by the

Criminal Investigation Department (CID) of the Indian Police. During this period, she was detained in solitary confinement and offered inducements, such as the chance to study overseas, if she would betray the struggle. However, she opted to stay quiet and asked the High Court judge if she was bound to answer the questions during her trials. When the court verified that she was not required to answer the questions, she announced that she would not answer any of them, not even to defend herself. The petite but determined woman stood firm on her decision. Following the verdict, she was sentenced to four years in prison (1942 to 1946). Additionally, two of her associates were convicted. Usha was imprisoned in Pune's Yeravda Jail. Her health deteriorated, and she was transferred to Sir J. J. Hospital in Bombay for treatment. Three to four police officers kept a constant watch on her in the hospital to prevent her from leaving. She was returned to Yeravda Jail after her health improved. She was liberated in March 1946, the first political prisoner to be released in Bombay, on the orders of the interim government's home minister, Morarji Desai.

After her release, she got a PhD and taught for 30 years at Bombay University's Wilson College. In 1998, she was awarded Padma Vibhushan, one of India's highest civilian honours. As a student, a research assistant, a lecturer, a professor, and lastly as the director of the department of civics and politics, she had a lengthy history with Mumbai University. In 1980, she retired from the University of Bombay. She also played a significant part in the administration of Mani Bhavan, supervising its upkeep and image exhibits and holding discussions on Gandhi's legacy and the liberation movement. She produced numerous articles, essays, and books in both English and her native Gujarati. She was elected president of Gandhi Smarak Nidhi, an organisation devoted to preserving Gandhian heritage. The Nidhi acquired Mani Bhavan in Mumbai, the residence of Sardar Patel's daughter Maniben Patel, and converted it into a Gandhi memorial. She was the president of the Gandhi Peace Foundation. She was also actively involved in Bharatiya Vidya Bhavan's affairs. The Government of India affiliated her with several India's independence events.

Even though she was sick with a fever in August 2000, Usha attended the August Kranti Maidan anniversary festivities for the Quit India Movement, as she did every year. She arrived home fatigued and frail. Two days later, on 11 August 2000 at the age of 80, she passed away quietly. The fiery young woman who was instrumental in establishing an underground radio evolved into the calm middle-aged woman who maintained the Gandhi memorial in the city that had been so significant to his movement.

In order to become a strong individual, one needs to listen to their own instincts and make their own decisions that would lead to their life fulfilling their purpose. Usha Mehta taught us that to diminish your doubts and your fears, you must have perseverance and confidence in yourself.

SATYAVATI DEVI
(1905–2010)

22

SATYAVATI DEVI

Throwing thunder through the Chimney

Satyavati Devi was the oldest freedom fighter and was influenced by Devki Devi, her aunt. Devki Devi was a widow who defied social conventions by continuing to study. She taught at Jalandhar's Kanya Mahavidyalaya. Satyavati too was admitted to Kanya Mahavidyalaya with encouragement from her grandfather. Her education shaped her approach toward life, country and relationships. At the time of her death on 26 October 2010, she was India's oldest living freedom fighter. She was imprisoned during the Quit India campaign and picketed liquor stores in response to Gandhi's request. Satyavati Devi continued to spin the charkha (spinning wheel) daily from her bed and sewed her clothes and her kid's clothes till she passed away.

Satyavati Devi was an Indian freedom fighter and a Gandhian. She attended Kanya Maha Vidyalaya in Jalandhar. In 1925, she married Lala Achint Ram. He was widely recognised as the Gandhi of Punjab and served in the Constituent Assembly and as a Congress Member of Parliament till his death in 1961. Her marriage was dowry-free, and she didn't wear a veil, as it was a condition Achint

Ram stipulated for marriage. She was affectionately referred to as Bijji or Mataji. She had a son Krishan Kant, India's former Vice President, and two daughters named Nirmala and Subhadra.

While her husband was imprisoned and was in and out of jail in Lahore, she continued to aid freedom fighters. She frequently brought food to Bhagat Singh, who attended the college across the street and visited the library at Lajpat Bhavan. Chandrashekhar Azad, the revolutionary leader, stayed at her house for three days before fleeing to Lahore.

On 26 August 1942, she and her two children, Nirmala, five, and Subhadra, thirteen, were arrested for their alleged involvement in the Indian liberation struggle. Achint Ram had already been imprisoned and Krishan Kant was jailed in the area during the Quit India campaign. In December, she requested her friend Karuna Chopra, who visited them in jail, to bring her a Tricolor, which Karuna smuggled in.

On January 26, 1943, Satyavati Devi along with other women prisoners pushed the Indian Tricolor up the chimney to the top of the British run Lahore Jail, and twenty of them began singing a patriotic hymn. As punishment, several of the women were sent to other jails, and she was thrown alone in a room. According to her account, "But as it turned out, that was the only room which had a light and I was able to read the works of Gandhi and Tilak and the Bhagwad Gita and spin the charkha. So, it wasn't a punishment." In prison, she organised a satyagraha and protested against the state of political prisoners' barracks.

The family remained in Lahore during Partition while Achint Ram assisted individuals on both sides in crossing across, and she came to Delhi only in October after Achint

Ram was elected to the Constituent Assembly. He was afterwards elected to the Parliament from Hissar and Patiala.

Following India's independence, she joined Vinoba Bhave's Bhoodan Movement alongside her husband Lalá Achint Ram. Both pleaded with landowners to make their land available to landless labourers, for this cause during Bhoodan Movement Satyawati walked village to village in North India. She was always well-respected by everyone who visited her home. Her son Krishna Kant was appointed Governor of Andhra Pradesh in 1989 and served until 1997 when he was appointed Vice President. Krishna Kant entertained his mother at his home occasionally. When her son died in 2002, she sat alongside him till his body was moved for cremation. For eight years, she outlived him and many of her younger coworkers, dying on 26 October 2010 at the age of 105. The following day, she was cremated in her native village with full state honour. A year before her death on 9 August 2009, she was honoured by then-President Pratibha Patil as part of the Quit India Movement's 67th anniversary commemorations.

At the time of her death on 26 October 2010, she was India's oldest living freedom fighter. On her 100th birthday, Prime Minister Manmohan Singh placed a blanket around the brave liberation warrior's shoulders and presented her with a commemorative stamp in the remembrance of her late son, Krishan Kant.

Satyavati Devi's life experiences teaches us to accept and adopt the changes that largely accomplishes the good of many. While being open-minded about the societal expectations continue on the line to be inspiring and daring. For this beautiful human being the age was just a number.

SUCHETA KRIPLANI
(1908–1974)

23

SUCHETA KRIPLANI

Specs beyond Horizons

Sucheta Kriplani was an important member of the freedom struggle with Mahatma Gandhi and a fervent advocate for increased female participation in politics. However, few are aware that Sucheta Kriplani was elected as India's first female chief minister in the state of Uttar Pradesh. Moreover, she was one of the 15 women selected to create the Indian Constitution as part of the legendary Constituent Assembly. Due to her anti-British political efforts, and participation in Quit India Movement, she was imprisoned for one year in 1942.

She was born in Ambala, Haryana on June 25, 1908, to a Bengali Brahmo family. Her father Surendranath Majumdar was a medical officer, a position that required frequent relocations. Her last degree was a Master of Arts in History from St. Stephen's College in Delhi. This was a time when nationalist emotions pervaded the country's atmosphere and the freedom struggle gained strength. She was not endowed with an iron determination and excellent leadership qualities at birth. According to her autobiography, An Unfinished

Autobiography, she was a shy youngster who was self-conscious about her beauty and intelligence. The period in which she grew up and the circumstances she encountered formed her personality. Sucheta and Sulekha were both eager to join India's budding Independence movement. In her book, Sucheta recounts one episode that is quite captivating. After the massacre at Jallianwala Bagh, the Prince of Wales visited Delhi. Her school's female students were transported to the Kudsia Garden to honour the Prince of Wales. Both sisters wanted to refuse, but were unable to; this made them furious at their apparent timidity. She points out in her book, An Unfinished Autobiography, "I could understand enough to feel great anger against the British (after hearing about the Jallianwala Bagh massacre). We (Sucheta and her sister Sulekha) vented out anger on some of the Anglo-Indian children who played with us, calling them all kinds of names," she writes.

Sucheta, who had just turned 21 and graduated from college, desired to enter the freedom movement. Unfortunately, her ambitions were delayed for a few years with the deaths of her father and sister in 1929, having left her with the responsibility of caring for a big family. She became a professor at Banaras Hindu University instead. Despite this, she continued to preach the message of the independence struggle. According to historical reports, instead of taking classes, she was discussing the significance of the independence movement with her students. She met her future husband, another revered leader of the liberation movement, Jivatram Bhagwandas (Acharya) Kriplani, twenty years her elder, at the university. In addition to being educational institutions, these universities also produced the ground warriors of the Independence cause.

Acharya Kriplani would frequently visit BHU in search of recruits to join India's freedom cause. Acharya Kriplani and Sucheta got to know each other well following the 1934 Bihar earthquake, during which they both participated in relief efforts for the victims. During this time, Jamnalal Bajaj urged her to join his Mahila Ashram in Wardha, an institute that intended to recruit women for the freedom movement. Even when she planned to marry Acharya Kriplani in 1938, their families and Mahatma Gandhi opposed the union. Despite her respect and admiration for Gandhi, Sucheta was determined to marry Acharya. Gandhi felt that the marriage would deprive him of Acharya's contributions to the liberation movement. To Gandhi's this fear, she replied that he would gain two workers for the freedom struggle. Eventually, the couple got married.

Kriplani joined the liberation struggle movement a few years after her marriage. In 1939, she left Banaras Hindu University to join Acharya in Allahabad. During this period, she became involved in the Indian political movement and joined the Congress party. She was entrusted with key tasks by the All-India Congress Committee; she was in charge of the Foreign Department and the Women's division. She worked diligently from 1940 to 1946 to organise the Women's section of the Congress Party. She created the All-India Mahila Congress, the women's wing of the Indian National Congress. Her anti-British political efforts earned her a one-year prison term for her participation in the 1942 Quit India Movement. She also assisted Gandhi closely throughout the

Satyagraha. In an interview, she discusses her experience with this mission, including the obstacles she had to overcome to promote women's participation in the Indian political and independence movement. She noted two major issues: the possibility of imprisonment associated with involvement in the freedom movement and the party's unwillingness to commit sufficient finances to the Women's section due to a shortage of financial resources. She was able to mobilise women to engage in the fight for independence amid these obstacles. She joined the subcommittee that drafted the charter for the Indian constitution. During the Independence Session of the Constituent Assembly on August 14, 1947, she performed Vande Mataram a few minutes before Nehru made his famous "Test of Destiny" address. Additionally, she was the founder of the All-India Mahila Congress.

Acharya and Kriplani were active Congress Party members. Acharya left the Congress in 1950 due to his disputes with Nehru and founded the Kisan Mazdoor Praja Party (KMPP). Kripalani was a member of KMPP for several years until rejoining Congress in 1952. She won the New Delhi Parliamentary seat on a KMPP ticket in 1952, but quit her husband's party shortly thereafter due to political disputes and rejoined the Congress. She won the same seat in 1957, but this time on a Congress ticket. "Both husband and wife had different political loyalties and never questioned each other, exemplifying democracy of a different kind." Kriplani was an experienced politician who was involved in the Uttar Pradesh political arena. Under the leadership of Chandra Bhanu Gupta as Chief Minister in

1960, she was appointed Minister for Labour, Community Development, and Industries.

Meanwhile, she had also become a member of the Uttar Pradesh Legislative Assembly. From 1960 to 1963, she served as Minister of Labour, Community Development and Industry in the UP government. In October 1963, she became the Chief Minister of Uttar Pradesh, the first woman to hold that position in any Indian state. She was elected one last time to the Lok Sabha in 1967, from the Gonda constituency in Uttar Pradesh.

She retired from politics in 1971 and remained in seclusion till her death in 1974. Given the small number of women in politics at the time, her achievements were remarkable, and she retained a reputation as a tough administrator.

We should all be courageous and be prepared to challenge the orthodox if we believe otherwise. Sucheta Kriplani shows us the power of her voice at the political and even at the household where her husband Acharya Kriplani to have different political loyalties. She continued to exemplify democracy of a different kind.

TARA RANI SRIVASTAVA

24

TARA RANI SRIVASTAVA

Life from Bangles to Bullet

Marching of a young bride Tara Rani Srivastava with the Tricolor in hand along with her husband towards Siwan Police Station in 1942, doing unthinkable when her husband was shot, not only bandaged her husband's wounds but continued to march as freedom of her country was the sole motive of her life. Several commoners battled against the British Raj and even gave their lives to ensure the independence and prosperity of their land. These people had modest origins and often lacked formal schooling. Some of them even fought against the customs of the time, which prohibited women from leaving their homes. Tara Rani Srivastava was one such person who came from a humble background. She was an Indian freedom fighter who participated in the Quit India Movement.

Tara Rani, who was born into a humble family in the Bihar district of Saran, close to Patna, was motivated by the unfolding struggle for independence at a very young age. She married Phulendu Babu, who was already involved in the freedom struggle. Her marriage did not deter her patriotism. Her spouse was an active member of the Quit India Movement led by Mahatma Gandhi. Tara Rani's patriotic enthusiasm increased with each passing day following her marriage, and she had little regard for the conventional society in which she resided. She organised like-minded women to educate them on Gandhiji's independence movement at a time when women did not leave the house. She mobilised the women of her nearby villages and inspired them to stand up in opposition to British rule. Tara Rani motivated rural women to engage in the Quit India Movement by serving as a source of strength for them.

An incident showed her remarkable bravery and unwavering patriotism, transforming her from an ordinary woman to a freedom fighter who deserves credit. On August 12, 1942, Mahatama Gandhi called out to raise the national Flag atop the police station in Siwan. Phulendu Babu gathered a large gathering of men and women to raise the national Flag, with his wife Tara Rani Srivastava. The pair, not long after their wedding, stood up against the British administration and strode forward, brimming with love for their country. By raising the national Flag on the roof of the police station, they intended to emphasise the strength of a united India against the British. They marched towards the Saran police station while reciting anti-British Raj slogans, with Tara Rani carrying the Tricolor and a massive crowd of men and women following them. In an attempt to control the rally, the police lathi-charge on the demonstrating crowd. But as a result of the administration's harshness, the unrest grew, and when the demonstrators became unmanageable, the police opened fire on them.

Phulendu Babu was injured by the police bullets; however, Tara Rani was fearless. She not only treated her husband's injury but also proceeded to march toward the police station while carrying the national Flag. Due to her devotion to the independence struggle, she made the liberation of her country her entire purpose in life. By the time she returned, her husband had succumbed to the police bullets. She realised that her husband had given his life for the greater cause of the country. As many would have anticipated of a woman of her day, her patriotism did not perish with her husband. Instead, her resolve grew even stronger. She was not the one to be restrained by the conservative society's traditions and norms. She knew the trials of widowhood, nevertheless, she remained active in the freedom struggle until August 15, 1947.

During a time when women were deprived of fundamental rights and chose to remain inside their homes, Tara Rani inspired and motivated women to join the freedom movement and helped them recognise the significance of their role. From a woman's perspective, her efforts to the cause of liberation were commendable. Carrying the Tricolor as a young woman, while leaving her wounded husband on the ground exemplified admirable dedication to the freedom movement.

Tara Rani bestowed to us that strong determination to do something needs courage but displaying such a commendable resolve and continuing to work for their country's freedom after facing loss of her loved one needs extraordinary valor.

PRABHAVATI DEVI
(1906–1973)

25

PRABHAVATI DEVI

Invisible Wounds

As a committed follower of Mahatma Gandhi, Prabhavati Devi spent most of her youth in the Gandhi Ashram. She took part in the 1932 Civil Disobedience campaign and was arrested for her participation. Her contribution to the Quit India Movement was noteworthy. She was caught violating the Salt Law and confined to Bhagalpur prison. In Patna, she founded the Mahila Charkha Samiti to continue the Mahatma's spinning wheel work.

Prabhavati Devi Narayan was an Indian independence activist and the daughter of a famous Lawyer Brajkishor Prasad and Phool Devi in Srinagar in the present-day district of Siwan in Bihar.

She was married to Jayaprakash Narayan. Jayaprakash Narayan travelled to the United States to initially study science in California but instead enrolled at the University of Wisconsin to study Marxism. She moved to Gandhi's ashram and dedicated herself entirely to Kasturba Gandhi, who began to view Prabhavati as her daughter. She also became Kamala Nehru's confidante and developed a close relationship with her. She was arrested following a demand for a boycott of foreign goods in Lucknow in 1932. Gandhi Ji and Rajendra Prasad entrusted her with the task of organising girl volunteers. She was imprisoned and transferred to Bhagalpur jail when the Quit India Movement began.

When her husband returned, he was seen as a rebel, which caused her to have frequent disagreements with him due to her Gandhian beliefs. Despite this, they respected one another and agreed not to have children until the country was free of foreign rule. Her husband, by now a cult figure in India, joined the Sarvodaya movement and actively participated in peace overtures in North East India and the Middle East under her influence. She founded the Mahila Charkha Samiti in Patna to engage deserted women in the Gandhian charkha or spinning wheel movement.

Her last several years were especially difficult because she was diagnosed with advanced cancer. She passed away on April 15, 1973.

Sometimes, the Motherland demands your highest sacrifice, even when you choose to prevent your own motherhood. Prabhavati Devi displayed her true self and left us with a perception that collective welfare is greater than self.

PRITILATA WADDEDAR
(1911–1932)

26

PRITILATA WADDEDAR

Don't treat her as a bird—she is far from being ensnared

Pritilata Waddedar was one of those revolutionaries who were wounded in battle and decided to end their lives by swallowing cyanide rather than be apprehended by the British. She was a fearless warrior, as evidenced by the fact that she was Bengal's first women martyr. Pritilata took up arms against the British and even led various anti-British actions under Master Da's command. While each of her activities during this period was remarkable for a woman revolutionary, nothing compares to the day Master Da chose her to lead an expedition to revenge on the Jalalabad atrocity by torching the Pahartali European Club which had a sign board that read "Dogs and Indians not allowed". Pritilata Waddedar is not a forgotten revolutionary. However, she is little known even in her native Bengal, where she spent her brief years vigorously resisting British control.

Pritilata was born in the village that witnessed the historic armory attack in 1930. She came from a middle-class family;

her father worked as a clerk. She was one of six siblings and whose resolve to fight for the abolition of British dominance sprang from contemporary resistance movements. Waddedar was born in Chittagong, now in Bangladesh, and was a bright student, having spent her school years in her homeland. During her time as a student at Eden College in Dhaka college, Wadderdar's anti-British sentiments emerged gradually, when she made contact with other women organising semi-revolutionary organisations. One such meeting occurred with Leela Nag, a student at Dhaka University and a friend of Subhash Chandra Bose, who created the Deepali Sangha, a revolutionary organisation that provided women with combat training. Waddedar moved to Calcutta to pursue higher education and enrolled as a Philosophy student at Bethune College, which is affiliated with the University of Calcutta.

Pritilata Waddedar joined nationalist Surya Sen's group to fight the British after a brief spell as a school teacher. She was among the early members of the armed independence movement led by prominent Bengali revolutionary Surya Sen or lovingly referred to as Master Da, whose name will forever be associated with the infamous 'Chittagong armory raid' as its chief conspirator. She expressed her desire to join his armed squad of young rebels. Although Master Da was initially skeptical, he was won over by her determination to topple the British, and he admitted her as a female comrade in his underground group. According to several accounts from the 1930s, members of Sen's group originally opposed her enrollment but eventually succumbed after discovering her commitment to the cause of the country's independence and her ability to carry out missions unseen by the authorities

Waddedar, aged 20, conspired with Surya Sen, Ganesh Ghosh, Lokenath Bal, Ambika Chakrabarty, Anand Prasad Gupta, Tripura Sen, Bidhubhusan Bhattacharya, Kalpana Dutta, Himangshu Sen, Binod Bihari Chowdhury, Subodh Roy, and Monoranjan Bhattacharya, among others, during the Chittagong Armory Raid. While the party was unable to find the armory, they did manage to destroy the telegraph and telephone cables. This did not go well with the British authorities, who managed to locate the runaways four days later in the Jalalabad hills near Chittagong, following a widespread raid on the fugitives. With several thousand troops closing on them, the rebels had little hope—12 were murdered in the ensuing gunfight, while the remainder, including Sen, reorganised and dispersed.

While each of her activities during this period was groundbreaking for a woman revolutionary, nothing compares to the day Master Da chose her to lead an expedition of 40 men to revenge for the Jalalabad atrocity by torching the Pahartali European club in 1932. While the loss of fellow compatriots in an exceedingly outnumbered encounter was a motivating factor for revenge, this specific club was singled out for its supremacist signboards that read, 'Dogs and Indians not allowed.'

A party of ten was trained in the use of guns and taught how to swallow potassium cyanide if the necessity arose under Waddedar's leadership. Pritilata dressed up as a Sardar and was unidentifiable, while her team wore shirts and lungis as they laid siege to the European club and set it on fire. Unfortunately, their victory was fleeting, as British troops quickly retaliated. Following an intensive chase punctuated by gunfire from both sides, Pritilata and her party were ambushed by soldiers in a brutal fight. She suffered a bullet wound, rendering her unable to flee the scene with the rest of her company. When she realised that yielding to the British was the only way out, she devised a ruse that would create an opportunity for her comrades to escape. Her attachment to her nation was so strong that she chose suicide over surrendering and giving out her faction to the British. Pritilata Waddedar evaded capture in those conditions by ingesting cyanide. Even in her death, she sent a strong message to the British, as a true rebel. She was just 21 years old at that time.

Waddedar's graduation degree had been revoked by the British administration of Bethune College, which is affiliated with Calcutta University. However, almost eight decades after her death, the University of Calcutta posthumously awarded Waddedar her overdue Bachelor of Arts degree with Distinction for the year 1932. It is learnt that a copy of Waddedar's graduation diploma and mark sheets were presented to the Birkanya Pritilata Trust in May 2018, which was founded in her honour and is based in Waddedar's native village, Dhalghat, Patiya, Chittagong. There is a statue of Pritilata Waddedar in Kolkata's Maidan area, along with some other freedom fighters, and it is the only one in the country. Waddedar is depicted in the towering statue dressed in a khadi sari with sharp folds, with one arm outstretched and the other balled up in a fist, perhaps signaling her commitment to the cause of liberation. Her fearlessness, strong will and refusal to give in to British pressure deserve to be honoured more than it has been.

Listening to your instincts and then sailing against the waves requires a great level of determination and passion. Pritilata Waddedar boldly declared herself as a Revolutionary when crushing patriotism fervor was on peak, she was willing to go against any odds for the freedom of our motherland, however hard or fearful it was to be.

ARUNA ASAF ALI
(1909–1996)

27

ARUNA ASAF ALI

The Heroine of 1942

Aruna Asaf Ali was popularly known as the "Grand Old Lady of the Indian Independence" for her role in the freedom struggle. She fought for independence fearlessly, enduring jail and escaping underground to continue her campaign despite several arrest warrants. The British had even offered 5000 rupees prize for her capture. Known for her independence, she even refused Gandhi's plea in 1946 to surrender herself: "I have been filled with admiration for your courage and heroism. You are reduced to a skeleton. Come out and surrender yourself and win the prize offered for your arrest...", he had written. Following her marriage, Aruna became interested in the independence movement. In 1930, she was arrested for addressing meetings and processions as part of the Salt Satyagraha but was released following great public outrage. In 1932, she was arrested again and went on a hunger strike in jail in protest of the way political prisoners were treated. Aruna Asaf Ali was an Indian independence activist, a freedom fighter, and one of the most prominent female personalities in India's liberation struggle. She was a revolutionary who assumed leadership during the Quit India Movement (1942).

She was born on 16 July 1909 at Kalka, British Punjab (now a part of the Haryana State) to Upendranath Ganguly and Ambalika Devi. Ambalika Devi was the daughter of Trailokyanath Sanyal, a prominent leader of the Brahmo Samaj (a monotheistic reformist organisation that emerged during the Bengal Renaissance). Upendranath and Ambalika Devi were both fervent supporters of the Brahmo Samaj too and were quite liberal. Aruna lived a typical childhood in a liberal, upper-caste Bengali family. Aruna Ganguly was extremely well educated. She was the eldest child of the family.

She studied at Lahore's Sacred Heart Convent. While in school, she fell in love with Catholicism and decided to become a Roman nun. Her family was so enraged that they enrolled her at a protestant school in Nainital. Aruna completed her graduation, at the All-Saints College in Nanital. While working as a teacher at the Gokhale Memorial School in Calcutta, she met her future husband, Asaf Ali. Asaf Ali was an eminent member of the Indian National Congress (Founded on December 28, 1885). Despite familial opposition, because of the huge age gap of 20 years and different religions, she acted on her desires, and they got married in 1928. She became an active participant in the movement for independence and her association with the Indian National Congress began after she met Asaf Ali.

She shifted her focus to Indian politics in the hope of making a significant contribution. Gandhi's goals and beliefs, as well as the views of others in the Indian National Congress, had a significant influence on her. She became an active participant in the freedom struggle. Her first foray into politics began in 1930 when she actively participated in public protest marches during the Salt Satyagraha. In 1930, she was arrested for addressing meetings and processions as part of the Salt Satyagraha but was freed following great public outrage. In 1931, the Gandhi-Irwin Pact was signed, promising the freedom of all those incarcerated during the Salt Satyagraha, but not Aruna Asaf Ali. Only a concerted effort by other women freedom fighters and Mahatma Gandhi contributed to her release.

In 1932, she was arrested and imprisoned in Delhi's Tihar Jail for her participation in the liberation movement. She was released in 1933. While in imprisonment, instead of suffering her confinement and awaiting release, she organised political prisoners and protested against the inhumane treatment that inmates were receiving by going on hunger strike. Even though her protest resulted in an improvement in jail circumstances, Aruna was transferred to solitary imprisonment in Ambala and remained politically inactive for ten years following her release, until the Quit India Movement.

Following the adoption of the Quit India resolution by the All-India Congress Committee on August 8, 1942, the British government detained all of its major leaders, including Mahatma Gandhi and Jawaharlal Nehru. On the 9th of August, to prevent the revolutionary spirit from dying, she seized control of the remaining session and raced to the front lines of battle. Responding to Gandhi's call to "do or perish," she defied the British on 9 August 1942 by hoisting the Tricolor at Bombay's (now Mumbai) Gowalia Tank Maidan (now Azad Maidan), providing the movement one of its most memorable symbols. Aruna was hailed as the "Heroine of the 1942 agitation." The police fired upon the crowd but Aruna stood her ground in the face of danger. Troubled by her open rebellion, the police attacked the meeting, hurling tear gas at the crowd and crushing the Flag she had raised. However, the harm was done, as demonstrations and rallies erupted across the country. The lack of senior leadership did little to stem the tide of nationalism.

She relocated from Bombay to Delhi to organise the resistance effort. However, alarmed of being apprehended by the authorities, she fled underground, evading arrest. Aruna edited the Congress party's monthly magazine 'Inquilab' while she was underground. She also utilised underground radio and pamphlets to keep up the fight while she was in hiding. During this time, she had close encounters with socialism and its notable leaders, including Jayaprakash Narayan, Ram Manohar Lohia, and Edatata Narayanan.

She emerged from hiding in 1946 when the warrant against her was eventually dismissed. Aruna joined the Communist Party of India upon independence and was instrumental in establishing the CPI's women's section, the National Federation of Indian Women, in 1954. This outstanding woman was elected Delhi's first Mayor in 1958. While Asaf Ali served as Minister of Communication following India's independence, she tried to improve women's status. She advocated for women's education, believing that it was the only path to emancipating women from the male-dominated society's confines. She accomplished this goal by founding the weekly periodical 'Link' and the daily newspaper 'Patriot'. In 1955, the Congress Socialist Party merged with the Communist Party of India, and she was elected to the Central Committee and Vice President of the All India Trade Union Congress. She did, however, leave the Communist Party in 1958. In 1964, she rejoined the Congress Party but did not engage in active political activity. Despite her opposition to the Emergency, she stayed close to Indira Gandhi.

Aruna was committed to mobilising support for Dalits and was opposed to "necessary industrialisation," which she claimed would result in environmental damage and social estrangement. She strove to advance women's rights, she believed that education and improved healthcare would benefit women more than reservation ever could.

Aruna was lauded by both nationalists and leftists. She received the Padma Vibhushan in 1992 and the Bharat Ratna posthumously in 1997 for her contribution to India's freedom struggle. The All-India Minorities Front annually bestows the Dr Aruna Asaf Ali Sadbhawan Award in her honour. Throughout her career, she garnered numerous honours. Though she avoided publicly accepting decorations, she received the Order of Lenin and the Lenin Peace Prize in 1965; in 1991, she received the Jawaharlal Nehru Award for International Understanding; in 1992. The Government of India commemorated her service to the Indian National Congress and the liberation struggle with a stamp in 1998.

This brave freedom fighter died on 29 July 1996. Her contribution to the struggle for independence and the national movement is immeasurable. Aruna Asaf Ali had this inner strength with which she resisted British control and the revolutionary fire and worked for the betterment of women, the impoverished and the oppressed.

If we analyze Aruna Asaf Ali's life we see that she was a woman who was able to stand up for herself and for others too. She acted on her own wishes and didn't allow religious or societal conventions to stop her to following her heart, while she was as fearless in her fight for independence, endued imprisonment and remained underground to continue her fight.

DURGAWATI DEVI
(1907–1999)

28

DURGAWATI DEVI

She's not Frail—She's Fire

Durgawati Devi commonly referred to as "The Agni of India," is still remembered as a hazard to the British police. She had a significant impact on revolutionaries such as Bhagat Singh, Ashafaqullah, and Chandra Shekhar Azad. She was an ardent member of the Naujawan Bharat Sabha, and she helped Bhagat Singh escape from Lahore following the John P. Saunders murder (1928). Durgawati Devi was one of the few women revolutionaries who actively participated in the armed revolution against the British Raj, and other members of Hindustan Socialist Republican Association (HSRA) affectionately called Bhabhi (elder brother's wife) and was also popular as Durga Bhabhi.

Durgawati Devi was born on 7 October 1907 in Allahabad, Uttar Pradesh, to a wealthy family from the modern-day district of Kaushambi near Prayagraj. Durgawati Devi was the only child of a Brahmin couple from Gujarat. Her mother passed away when she was young, and her father left her with her aunt to raise. She got married to Bhagwati Charan Vohra and met the revolutionaries in Lahore through her husband, another Hindustan Socialist Republican Association (HSRA) member Bhagwati Charan Vohra, the son of a rich Gujarati, Shiv Charan Das, who worked for the railroads and was given the title of Rai Sahib. Bhagwati Charan attended National College, Lahore, where he met Bhagat Singh, who, along with Yashpal and Sukhdev, became frequent visitors to the family home.

Bhagwati Charan was able to devote a substantial amount of time and resources to social and political action. In late 1928, he and Durgawati Devi joined the party, and he became one of the most prominent HSRA ideologues. Durgawati Devi had also donated her collection of jewels and cash to the HSRA so that it could purchase rifles and ammunition for the rebels. Later, the organisation got strongly affiliated with her as well. In 1925, Durgawati Devi gave birth to her son Sachinanda, but she stayed dedicated to teaching and continued to work at a women's college in Lahore until she was driven to go undercover in 1929.

Early in December, Bhagwati Charan departed Lahore to attend the annual convention of the Indian National Congress in Calcutta. Durgawati Devi was aware of the murder of John P. Saunders, a British Officer in Lahore who, on the instructions of James Scott killed Lala Lajpat Rai during a lathi charge, as the entire city of Lahore was abuzz with the news when HSRA posters proudly claimed credit. Consistent with the party procedure, she did not ask any questions, despite knowing that Bhagat Singh was involved. On December 19, 1928, two days after killing John P. Saunders, Sukhdev Thapar contacted Durgawati Devi for assistance, which she agreed to provide. On the evening of

19 December 1928, when Singh and Sukhdev visited her home, Sukhdev introduced Singh as a new friend. Durgawati Devi was unable to recognise Singh. Then, Sukhdev informed her the truth, stating that if she would not identify Singh in his new, clean-shaven look despite knowing him well, then neither would the police, who would be searching for a bearded Sikh. They decided to take the early morning train from Lahore to Bathinda on the way to Howrah (Calcutta) the following day.

Against societal customs of the time that restricted interaction between unmarried men and women, Durgawati Devi decided to pose as his wife to assist him to escape Lahore. Bhagat Singh had shaved off his beard trimmed his hair short and donned in western clothing to avoid being recognised. Taking her son Sachi and escorted by Rajguru carrying their bags (who pretended to be the young family's servant), they slipped unnoticed under a police barricade and entered a first-class train carriage bound for Lucknow, where they changed trains on route to Calcutta.

Both men carried loaded revolvers in preparation for any unplanned event. They avoided arousing the police's suspicions and boarded the train. At Cawnpore (Kanpur), they transferred to a train bound for Lucknow, as the CID at the Howrah (Kolkata) railway station routinely scrutinizes people on the direct train from Lahore. Rajguru left Lucknow for Benares (Varanasi) while Bhagat Singh, Durgawati Devi, and the newborn travelled to Howrah. After attending several sessions of the Calcutta Congress, a few days later, Durgawati Devi returned to Lahore with her young kid. Azad also fled Lahore with the assistance of women. Azad pretended to accompany Sukhdev's mother and sister on a pilgrimage in a disguise. However, Durgawati Devi led Jatin Das' funeral procession from Lahore to Calcutta when he died on a 63-day hunger strike in jail. Along the entire route, large crowds joined the funeral procession.

In the meantime, the preparations were going on for the attack on the Legislative Assembly included Bhagwati Charan, who had learned how to create bombs from revolutionary groups in Calcutta. Durga Devi was called by her husband to Delhi in early April 1929 to bid farewell to Bhagat Singh. He proceeded directly to the Assembly, where he and BK Dutt set in motion the circumstances that led to his execution.

Durgawati Devi and her husband assisted Vimal Prasad Jain, a member of the HSRA, in operating a bomb factory on Qutub Road, Delhi, under the guise of "Himalayan Toilets" (a smokescreen to conceal the bomb-making goal). They handled dangerous bomb materials there. Her primary responsibility was to transport weapons and ammunition to revolutionaries. In the months that followed, law enforcement began to move in on the rebels. In Lahore, detectives uncovered the HSRA's bomb factory in Kashmir House, which was rented under Bhagwati Charan's name. Sukhdev, Jai Gopal, and Kishori Lal were arrested as a result. Bhagwati Charan was not present during the raid and thereafter went into hiding.

In 1929, Bhagwati Charan and Yashpal conspired and attempted to bomb the Viceroy's train. Thereafter, he joined Azad, Yashpal, Vaishampayan, Sukhdevraj, Sushila, and Durgawati Devi in a conspiracy to liberate Bhagat Singh from prison by attacking a police truck during his transportation from Borstal to Central Jail in Lahore.

But, on May 28, 1930, Bhagwati Charan passed away. He was testing a bomb in the forest near Ravi when it unexpectedly exploded in his fingers. His companions went for assistance, but by the time they returned, it was too late. Later, cops discovered him dead in a small grave nearby. After Durgawati Devi's husband was killed in a bomb testing operation at Lahore prison, many believed that her struggle had ended. Despite the loss of her spouse, she continued to struggle against British control.

Her subsequent actions and locations are unclear, other than the fact that she was continually on the move, remaining one step ahead of the cops. Bapat had driven Prithvi Singh and Durgawati Devi to their destination in the early evening. Sir Geoffrey de Montmorency, the Governor of Punjab, was their primary target; but, the protection around his residence in Malabar prevented them from approaching it. Disappointed, they decided to attack a police station instead and eventually saw the two Britishers. At midnight on 8 October 1930, she opened fire on a European couple in front of the police station in South Bombay. This was later described as one-of-a-kind shooting incident in which a woman played a prominent role. It turned out that the victims of the shooting were a police sergeant, Sergeant Taylor, and his wife. His hand was pierced by a bullet, while his wife's leg sustained three wounds. Durgawati Devi, Prithvi Singh, and Sukhdevraj conceived the action in a haste to avenge the death sentences of Bhagat Singh, Sukhdev, and Rajguru. They decided to kill Lord Hailey, the former governor of

Punjab and a sworn adversary of the rebels. Although the Governor escaped, his aides were injured. She was arrested and sentenced to three years in prison.

She was also important in organising protests and marches at the time seeking Singh's release. She voiced her opposition to the sentences handed down to Bhagat Singh, Sukhdev, and Rajguru. She contributed all her jewellery and cash to rescue Bhagat Singh and his trial-bound colleagues.

After Indian independence, unlike other freedom warriors, Durgawati Devi began living as a normal person in quiet seclusion in Ghaziabad. She started a school for impoverished children in Lucknow in 1935. This extraordinary courageous woman died in Ghaziabad on 15 October 1999 at the age of 92.

Soha Ali Khan, who portrayed her in Rakeysh Omprakash Mehra's 2006 film Rang De Basanti, made a brief allusion to her role.

We learn never to stop until the goal is achieved from Durgawati Devi's life. This fiery woman remains always one step ahead of the enemies. She showed extraordinary level of confidence, trust and courage in herself and her team, which is essential to achieve any target.

BINA DAS
(1911–1986)

29

BINA DAS

When love impels to shot fire

There have never been handcuffs awarded during a convocation ceremony, but for Bina Das the ceremony came with a twist. In the spring of 1932, Das discovered that the Governor of Bengal, Stanley Jackson, would attend convocation ceremony of Calcutta University. 21-year-old Bina Das opened fire in the Calcutta University convocation hall on Stanley Jackson. She was meant to receive her diploma on the same day in 1932 at the same place. Das was sentenced for nine years for an act motivated by a strong desire to see her native land liberated from British rule.

Bina Das was a nationalist and revolutionary from West Bengal. Das was a member of Chhatri Sangha, a semi-revolutionary organisation for women in Kolkata. She was born to Beni Madhab Das and Sarala Devi, who were highly interested in the Brahmo Samaj and the independence struggle. The couple believed in providing their children, particularly their girls, with freedom, education, and a thirst for knowledge. Rare characteristics in the early 1900s. In Calcutta, Das's mother Sarala Devi maintained a women's hostel called Punya Ashram that acted as a storage facility for the revolutionaries' bombs and weapons. Several residents of this hostel belonged to numerous underground revolutionary organisations. Bengali women's rights and the advancement of women were the emphasis of her parents' indirect participation in the independence movement. But as a lecturer, Beni Madhab Das inspired many of his pupils to fight for India's independence, most notably Subhash Chandra Bose.

Das writes in her autobiography, translated from Bengali by Dhira Dhar, how much "Subhash babu" was influenced by her father and how often he visited her parents' home. The memoir highlights Das's initial encounter with Bose. She recalls her mother telling Subhash about how her daughter was a huge fan of his. Bina Das was influenced by Bose's political views, which strengthened her anti-British stance. In one incident from her youth, when the wife of the British viceroy was asked to visit their school, Das objected to the concept of holding flower baskets that were to be scattered at their feet. She was humiliated and missed rehearsals, and was angered by the entire experience. Two more girls also walked out of the rehearsal, and joined Bina Das. Much perturbed with this incident small girls took a vow that they would sacrifice their lives for the freedom of motherland.

As she got older, she observed the invasion of her motherland, the loss of the freedom she craved, and the innumerable stories she heard of British brutality—all of which fueled a rage that could only be doused after their departure. During this period, she also joined Chhatri Sangha, a Calcutta-based, semi-revolutionary organisation led by women. With every word she spoke and action she took, she devoted herself to eliminating British citizens from the country. Bose continued to serve as a guide in Das' life, particularly after she enrolled at Calcutta University's Bethune College. The college library and publications advocating revolution and liberation bolstered Das's convictions and ambitions for an independent India. In 1928, Das and a group of fellow students organised the first student protests against the Simon Commission and were threatened by the college administration and the English principal if they did not apologize. The "overbearing Englishwoman" resigned from her position and left the university as a result of the students' protests against the Simon Commission and reluctance to adhere to the college's demands. This may be considered Das' first taste of success against British domination. This revolt provided the groundwork for what became known as the Chhatri Sangha, a women's student association whose activities were semi-revolutionary. Kalyani, Das's sister, was the secretary of this organisation. Members of the Chhatri Sangha were taught basic self-defense, including lathi lekha, where the women were instructed in the use of batons, similar to other revolutionary groups that arose in Bengal at the time. This student organisation recruited new members and was led by prominent revolutionaries such as Dinesh Majumdar.

Bose's involvement as a mentor and as an individual who profoundly affected Das is highlighted in her memoir. During their conversations and meetings when Das was a college student, the mentor and mentee reflected on their convictions and discussed the future of their cherished homeland. In the spring of 1932, Bina Das learned that the Governor of Bengal,

Stanley Jackson, would attend the convocation ceremony at Calcutta University. He was the guest of honour at the ceremony. Bina Das approached her friend Kamala Dasgupta, a rebel affiliated with the Jugantar, in order to obtain guns. Bina Das stated her that, "it would be a great occasion to register (her) protest against the empire". Dasgupta decided to provide Bina Das with the firearms she required to carry out her plans after a barrage of questions and answers, including pleas to inform Bina Das of the grave repercussions of opening fire on the governor. On 6th February 1932, Bina Das opened fire in the Calcutta University convocation hall on Stanley Jackson. Despite realising the severity of her conduct, she continued to fire five times at Jackson from close range until she was apprehended, disarmed, and arrested. Even though she was arrested, her name appeared on countless front pages of newspapers until the news spread internationally.

"I confess that I fired at the Governor on the last Convocation Day at the Senate House. I hold myself entirely responsible for it. My object was to die and if I had to die, I wanted to do it nobly, fighting against this despotic system... I fired at the Governor impelled by my love for my country which is being repressed," she firmly accepted her act during many interrogations following her arrest.

In 1939, following her early release, Das joined the Congress party. She was again imprisoned from 1942 until 1945 for her participation in the Quit India movement in 1942. She was a member of the Bengal Provincial Legislative Assembly from 1946 to 1947 and the West Bengal Legislative Assembly from 1947 to 1951.

In the early 1940s, her activism for the liberation struggle once again landed her in Presidency Prison, where she remained until her release in 1945. Throughout her continued struggle against the British, Das experienced more desolation and bloodshed in the lead-up to freedom. In 1947, she wed Jatish Chandra Bhaumik, a member of the Jugantar Party and a fellow rebel. The British authorities of Calcutta University refused Bina Das her degree because of her revolutionary actions, in a vain attempt to punish and pressure her for her anti-British position. In 2012, about 81 years after Das's death, the University of Calcutta issued her Bachelor of Arts degree with second class honours in English for the year 1931.

After India achieved independence, little is known of Das's life. In 1960, she was awarded the Padma Shri by the Indian government for her services to social work. According to other accounts, she died in misery and suffering in December 1986, and her body was found in a ditch in Rishikesh, so badly decomposed that it took officials days to identify her.

Each rebel must fight his or her own fights, but the great part is that the revolution never fails. Bina Das's existence mirrored the fire within her.

Executing Childhood vow to sacrifice the life for the freedom of motherland, shows us the passion Bina Das carried from her childhood. In her own words:

"Later in life I often remembered this childish vow, and in moments of weakness it gave me strength and hardened my resolve."

FREDA BEDI
(1911–1977)

30

FREDA BEDI

Love is deeper than labels

"It was my destiny to go to India," Freda stated. She considered herself Indian and clothed in the Indian style from the time she married B.P.L. Bedi. In 1929, she enrolled in one of Oxford's women's colleges, and by the end of her studies there, she had married an Indian, Baba Pyare Lal Bedi, and was prepared to identify as an Indian. Freda joined India's freedom struggle and was jailed for supporting the Indian national cause over Britain's. Freda was handpicked by Gandhi for individual satyagraha. She later became the first western woman to take full ordination in Tibetan Buddhism.

Freda Marie Houlston was born in the city of Derby in England's East Midlands, where her father ran a jewelry and watch repair business. Freda could only vaguely recall her father because he died in northern France when she was just seven years old. Freda's life was influenced by the tragic deaths of her father on the battlefields of northern France during the First World War, and of her second child when he was still a newborn. Her emotional sensitivity was noticeable from early childhood on. She also possessed a steely resolve. Freda transformed herself throughout her lifetime. In addition to her discipline and compassion, she had a restless, inquisitive nature. Already from the beginning, she was a woman of faith and spirituality, before adopting Tibetan Buddhism, the religion that came to define her.

At Oxford Freda became involved in politics. She started attending meetings of the Oxford Majlis where national-minded Indian students gathered. She also became close friends with rebellious young women at her college and attended meetings of the Labour Club and the communist October Club with them. Freda and her friend, Baba Pyare Lal Bedi (known to his friends as BPL (a sixteenth-generation descendant of Guru Nanak Dev, the founder of the Sikh religion. Baba Bedi had a pedigree of his own), met while they were both students at Oxford. In the early 1930s, cross-racial alliances were almost as uncommon as a girl from Freda's background gaining admission to a premier university. According to Freda, her engagement generated "a minor sensation" in Oxford. She believed she was the first undergraduate woman at Oxford to marry an Indian Sikh student.

Soon, Freda, her husband and their four-month-old baby, Ranga, embarked on a two-week voyage from Trieste, Italy, to the western Indian metropolis of Bombay via boat (now Mumbai). As a result of their student activism, the British government had already identified the pair as prospective revolutionaries and troublemakers. When they arrived in Bombay, their bags and cases were searched for seven hours, and "even Ranga's little napkin was taken off and searched," recalled Freda, "because they thought I might be carrying messages in it."

The next step was to adjust with the extended Indian family and Freda resolved to blend in. Pyare Lal Bedi's political stance included a refusal to inherit any of his family's fortune. They lived in a cluster of thatched houses without electricity or running water, in Lahore, the center of the freedom movement in Punjab and one of the main cities there. They kept chickens and a buffalo there. It could not have been the kind of life Freda anticipated, nor would she have been accustomed to living with her mother-in-law. Her mother-in-law was also a person with a big heart; despite her conservatism, she had welcomed a Christian into the family without hesitation. Freda mixed her role as a loyal daughter-in-law with political activism seamlessly, once the couple had settled in Lahore. To earn a living, Freda became a professor of English at Lahore's Fateh Chand College for Women. Together, they published Contemporary India, a distinguished quarterly journal. Freda also contributed a weekly column titled From a Woman's Window to The Tribune, the city's primary nationalist newspaper.

In the meantime, Pyare Lal Bedi enlisted as a satyagrahi, a truth-seeker, to resist wartime emergency authorities. When World War II broke out, the British imprisoned B.P.L., to prevent him from disrupting military recruitment. Freda determined that she, too, must take a stance, and, with Gandhi's support, she became a satyagrahi, openly disobeying the Emergency wartime laws. She travelled to the village of Dera Baba Nanak, the birthplace of her husband, and declared that she would break the law by urging the villagers not to support the military effort until

India was free. She was the fifty-seventh person in India to be handpicked by Mahatma Gandhi for individual Satyagraha. The Tribune, Punjab's leading Newspaper, put it on its front page: Mrs Freda Bedi left for Dera Baba Nanak where she will offer Satyagraha on 21st February." The British were determined to stop her, but they deemed it improper for an Indian police officer to arrest a British woman, in a handwoven sari who had declared she was going to lead a non-violent demonstration, satyagraha, so Administrators hastily dispatched an English police inspector Price from Amritsar to Dear Baba Nanak to arrest Freda.

The night before her demonstration, as she lay down to sleep, she felt vulnerable. Her husband was not there to comfort her, he had been imprisoned by the British for yet another agitation. Inspector Price came to arrest Freda at 4:00 a.m. before she could begin her Satyagraha. He looked everywhere, but he couldn't find her anywhere. It must have upset him greatly. He had been specifically sent by the British for this operation. Price posted his men around a hastily erected stage in the yellow mustard fields where she was to address the crowds. She had to be stopped before she got there. But in the morning Freda suddenly appeared on stage, "as if from nowhere", to the welcoming road of 40000 well-wishers. She had hidden herself; under the stage in the darkness of the night with her children. Before Price could reach her, she spoke to the crowd, telling them not to be violent, whatever the provocation. Violence would defeat their cause.

Inspector Price reached her with an unwieldy toupee shading his pink face. A bearded Indian Policeman spoke on his behalf: "Regretting it is my duty, but I must arrest you." She replied, "I am quite ready. Take me along with you." Freda was brought before a visiting magistrate that same morning—she has left her account of the trial:

It was finished in 15 minutes. The man on the other side of the table was quite young still and looked as though he had been to Oxford. His face was red.

"I find this as unpleasant as you do," he murmured.

Don't worry. I don't find it unpleasant at all."

"Do you want the privileges granted to an English woman?"

"Treat me as an Indian woman and I shall be quite content."

She was sentenced to six months in jail, which was fairly standard, and also to hard labour, which she regarded as vindictive. The Reuters news agency picked up the story of the "first English woman to join Mr Gandhi's passive resistance movement". It made headlines in her hometown, "Derby Wife of Indian sentenced." Mercifully Freda's sentence was cut short. An Indian Judge ruled that the "intention to commit a crime does not constitute a crime" since she was arrested before she began her march, the British had to let her out three months early. Before leaving, she persuaded her jailers to change the name of the "Lahore Female Jail" to "Lahore Women's Jail". Jails, she argued, jails had no gender.

By the time of Partition, the couple had another child, the international famed actor Kabir Bedi, and a few years later, a daughter, Gulhima. The family relocated to Kashmir and

Freda joined a left-leaning women's militia and collaborated with the extreme nationalists who came to power in Kashmir, where the Bedis' political significance lasted even after independence.

During an assignment for the United Nations in Burma in the 1950s, she discovered Buddhism for the first time and became a fervent convert. Her life moved in a new direction once more. When tens of thousands of Tibetans fled through the Himalayas on foot, she persuaded Nehru to deploy her to Assam to ensure that the refugees had adequate amenities. She went on to found a school for young lamas; she became a Tibetan Buddhist nun—possibly the first woman in Tibetan tradition to receive full ordination. Thirty years after she left Derby, girls from her old school there were raising funds for a new school Freda was establishing for young Tibetan lamas in the foothills of the Himalayas.

By 1966, she had abandoned her previous lifestyle to become a Buddhist nun. She was inducted by the 16th Karmapa Lama at Rumtek Monastery in Sikkim and given the name Gelongma Karma Kechog Palmo, becoming "possibly the first woman to receive this higher level of Tibetan tradition initiation." She convinced her mentor, the Karmapa Lama, to reach out to new devotees, accompanying him on a ground breaking five-month tour of the United States and Europe in 1974.

Freda passed away in Delhi in 1977. The fact that she delighted in defying tradition and confounding expectations is what makes her journey so captivating. Her closest Oxford friends remained in contact with her throughout her life, even when she received holy orders. Freda led a life that few of us could ever conceive of, as she defied boundaries of nationality, race, and how women should represent themselves in society.

We learn from Freda's life that listening to our inner voice and believe that each of us has a personal calling which is distinct and unique. When we channelise our energy to discover ourselves and know what love, we are able to offer love to others in the form of our service.

LAKSHMI SAHGAL
(1914–2012)

31

LAKSHMI SAHGAL

'Resilient' and 'Extraordinary' is an understatement

Lakshmi Sahgal was a passionate freedom fighter, a social activist who served in the Indian National Army under Subhash Chandra Bose (INA) and a doctor. Captain Lakshmi, as she was affectionately known, devoted her life to public service in a variety of capacities, including as the commander of an all-female regiment in the Indian National Army, a medical practitioner serving the poor and the disadvantaged, and a member of the Communist Party of India (Marxist). Sahgal adhered to the principles of socialism, anti-imperialism, equality, secularism, social justice, and women's empowerment in her personal and professional life.

Lakshmi Sahgal was born on 24 October 1914 in Madras to a family of a lawyer and a social activist, Sahgal was always brave and followed her heart. She began defying social norms and rigid beliefs at a very young age, speaking out against caste traditions. When untouchability was prevalent in India, she approached a tribal girl, held her hand, and invited her to play. She had to pay a price, but that never stopped her. Lakshmi witnessed the freedom movement alongside her mother, A.V. Ammukutty, a social activist and freedom fighter who subsequently joined the Constituent Assembly. Her father was a talented lawyer.

Lakshmi was inspired by her mother's work in the field of social work, inherited her father's brilliance, and went on to become a doctor. After completing high school in Madras (Chennai), she attended Madras Medical College, where she earned her MBBS in 1938. Lakshmi and her family were pulled into the ongoing independence movement over the interim years. She witnessed her mother's change from a Madras socialite to a fervent nationalist supporter, who one day marched into her daughter's room and took away all of her beautiful garments to burn in a bonfire of foreign products. Lakshmi was married to the pilot P.K.N. Rao, but she quickly realised it was a mistake and went to Madras to complete her MBBS. Lakshmi attended Queen Mary's College before deciding to study medicine and graduating with an MBBS from Madras Medical College in 1938. A year later, she earned her gynecology and obstetrics certification. She was employed as a physician in the Government Kasturba Gandhi Hospital in Triplicane, Chennai.

Lakshmi, however, moved to Singapore in 1940 after her marriage to pilot P.K.N. Rao failed and began her medical practice there, treating the Indian diaspora as well as Chinese and Malay patients. In addition, a year after receiving her diploma, she started a clinic in Singapore for the disadvantaged and Indian migrant labourers. There, she joined the Indian Independence League, a Singapore-based political organisation that prepared Indians residing outside India to claim independence from the oppressive British rule. It was established by experienced freedom fighter Rashbehari Bose.

In 1943, Subhash Chandra Bose assumed command of both the League and the INA, which had been founded in February 1942 by Captain Mohan Singh and other Indian prisoners of war. When the Japanese forces were defeated by the British Army in the Battle of Singapore in 1942, Dr. Lakshmi performed a pivotal role in caring for the wounded prisoners. A number of these inmates had not yet given up hope and desired to form an Indian Liberation Army. In July of 1943, Netaji Subhash Chandra Bose visited Singapore and Lakshmi learned of Subhash Chandra Bose's plan to form a women's battalion to fight for Indian freedom. Keen on sticking up for the country, Lakshmi requested a meeting with Bose and, with his assistance, established the Rani of Jhansi Regiment, one of the world's first women's contingents. This gave numerous women the opportunity to serve their country.

The transformation of Lakshmi Swaminathan into Captain Lakshmi marked the start of her extraordinary journey in the independence movement. Under her supervision, about 50,000 women were trained and participated in combat. She held this position in the women's army unit, the first female to do so on the entire Asian continent at that time. Her unit fought alongside the Axis powers against British soldiers. The hands that healed the wounds, the fingers that checked the pulse now had held the heaviest weapons, shot revolvers and threw grenades. In December 1944, the INA marched to Burma with the Japanese army, but in March 1945, as the tide of war turned against them, the INA leadership decided to retreat before entering Imphal. Captain Lakshmi was detained by the British in May 1945, and she remained in Burma until March

1946, when she was brought to India—at a time when the INA trials in Delhi heightened popular unhappiness with colonial rule and accelerated its downfall.

In 1947, she wed Colonel Prem Kumar Sahgal, with whom she served in the INA. In the years following India's independence, Lakshmi resumed her medical profession in Kanpur. In 1971, she joined CPM and remained an active member. With that, she entered the realm of policy-making and represented her party, the Communist Party of India (Marxist), in Rajya Sabha after independence. During the Bangladesh war of 1971, she travelled to Calcutta (as it was then known) and spent six weeks providing medical assistance to the displaced and the migrants in the border areas of Bongaon. She quietly kept serving people by treating thousands of poor patients for free.

Lakshmi Sahgal was one of the 1981 founding members of the All-India Democratic Women's Association (AIDWA). The organisation became a good platform for her to regularly discuss women's issues and launch multiple campaigns for the same. She travelled to Bhopal in 1984 with a medical team following the gas disaster. Her dedication to secular principles in the face of adversity was repeatedly demonstrated. During the anti-Sikh riots of 1984 in Kanpur, she fought crazed mobs to ensure the safety of Sikhs in and around her clinic. Sahgal was also imprisoned in 1996 for her participation in a campaign by AIDWA against the Miss World competition held in Bangalore 1996.

She stood against A.P.J. Abdul Kalam in the 2002 presidential elections as the joint candidate of the Left parties. Although she lost the election, she spent many days touring the country and delivering speeches at numerous rallies.

R.K. Narayan presented Dr Lakshmi Sahgal with the Padma Vibhushan, the second-highest civilian award, in 1998 for her outstanding achievements. Captain Lakshmi Sahgal International Airport is dedicated in her honour in Kanpur's Dehat district. She passed away on July 23, 2012, at the age of 97, after having a cardiac arrest.

Dr Lakshmi Sahgal was a real leader who stormed into the male-dominated army world. Sahgal frequently championed various causes, irrespective of the outcome. Her turbulent life spanned 65 years of the struggle against colonial control, the accomplishment of independence, and nation-building in India during the 20th and early 21st centuries. Captain Lakshmi remained steadfastly on the side of the poor and powerless throughout this period. Her selfless dedication to her patients and the underprivileged was incomparable and endearing to millions.

Lakshmi Sahgal is one such person who never let her passion to serve the country die. Sahgal saw it all, the British rule, India's freedom and the transformation of the nation over some time and she always followed her duty passionately. Her journey has been truly inspiring and there is so much that we can learn from it.

When you don't want to settle for less, you need to be unequivocally committed to your passion. One of the lion-hearted women, India ever had. Picked up the Gun of Indian independence. Her evolving as young medical student to the freedom struggler, as the leader of all woman Rani of Jhansi regiment of the INA, as a doctor and as of member of CPI (Marxist), Lakshmi Sahgal a woman of such a multifaceted personality.

RANI GAIDINLIU
(1915–1993)

32

RANI GAIDINLIU

The women who broke the shackles

She openly defied British rule, urging the residents of Zeliangrong not to pay taxes. Gaidinliu, or Rani Gaindinliu, also called the 'daughter of the hills,' spent 14 years in jail fighting for India's freedom. "We are free people, the white man should not rule over us...," preached the 13-year-old Gaidinliu to the people of her tribe in present-day Northeast India. She was a spiritual and political leader from Manipur's Rongmei tribe. She joined the liberation movement at the age of 13 and eventually commanded the socio-political campaign that drove the British out. At the age of sixteen, the courageous tribal girl, Rani Gaidinliu, was sentenced to life in jail for organising a movement to overthrow colonial authorities from Manipur and Nagaland. She was captured in 1932 and sentenced to life in prison. She was imprisoned for 14 years and liberated only after India became independent in 1947.

Rani Gaidinliu was born on 26 January 1915, in Nungkao village in Manipur's Tamenglong district. She was the fifth of Lothonang Pamei and Kachaklenliu's eight children, which included six sisters and a younger brother. The family was a member of the village's ruling clan. Her native village was Nungkao, which is the present-day Tamenglong district was a socially backward region. She did not receive a formal education due to a lack of schools in the area. In 1927, at the age of 13, Rani Gaidinliu became associated with freedom warrior and religious leader Haipou Jadonang, who was also her cousin. Haipou Jadonang was a local leader who had risen to prominence. The movement of Jadonang was a revival of tribal religion. It drew a large number of followers from the Zeliangrong Tribe (the Zeme, Liangmai and Rongmei). She joined the Heraka religious movement. Heraka was a social and religious movement that began in the 1920s in the Zeliangrong territory. Heraka, which means "pure," was a monotheistic religion whose adherents worshipped Tingkao Ragwang. It was started by Jadonang to oppose both the entrance of Christian missionaries as well as the British government's reforms. He witnessed the repression of the tribals at the hands of British officials, who forced them to work in harsh conditions and imposed a large annual revenue levy on each household.

Jadonang was seen as a major threat by the British to their power because of his immense popularity among the Zelianrong Nagas in many parts of Manipur. In 1931, he was arrested and hanged after a mock trial. In the meantime, Rani Gaidinliu organised her version of the Non-Cooperation Movement among the Naga tribes, obstructing British operations in the area. She led guerrilla attacks on the British authority. Gaidinliu was on the frontline of the Heraka Movement, taking up the mantle from her cousin to ensure that the movement did not perish with Jadonang. She drew a connection between her spiritual role as a leader of the socio-religious movement and her other identity as a patriot, echoing Mahatma Gandhi's national efforts to rally her people against the British.

She openly defied British rule, urging the residents of Zeliangrong not to pay taxes. She was aided in her fight against the British by donations from the indigenous Naga community. She gained the support of the majority of the population to refuse to collaborate with the British, and as a result, posed a serious threat to British control. To know her whereabouts, British police launched a manhunt for her. She travelled between villages from Assam to Nagaland, and Manipur to avoid being arrested by the police. Monetary prizes were announced for supplying information leading to her arrest.

Early in 1932, her men fought the Assam Rifles in armed conflict in the North Cachar Hills and Hangrum village. Fearful of her growing popularity and strong defiance, the British authorities ordered a special Assam Rifles contingent under Captain MacDonald to seize Gaidinliu. They got to know Gaidinliu's location in the village of Pulomi through sources. On October 17, 1932, the Captain deceived the rebels and sent his troop in the opposite direction, and Gaidinliu and her companions were lulled into a false sense of security. The British forces conducted a surprise attack on the village, and Gaidinliu and her followers were arrested without any resistance. She was handcuffed and carried on foot to Kohima, and later to Imphal for a trial. She was found guilty of murder and attacking the Assam Rifles. While most of her associates were either executed or imprisoned, she received a life sentence.

She served time at Guwahati, Shillong, Aizawl and Tura jail between 1933 and 1947. Many rebels proclaimed her and Jadonang to be their inspiration in refusing to pay taxes to the British. After the last of her followers, Dikeo and Ramjo was jailed in 1922, her movement began to decline.

On a visit to Manipur in 1937, Jawaharlal Nehru visited Gaidinliu in the Shillong jail and promised to pursue her release. Jawaharlal Nehru while acknowledging her role and contribution to the fight against the British called her the "Daughter of the Hills" and bestowed the title "Rani" or queen. He also wrote about her: "...and now she lies in some prison in Assam, wasting her bright young womanhood in dark cells and solitude. Six years she has been there. What suppression of spirit they have brought to her who in pride of her youth dared to challenge the Empire.... And India does not even know of this brave child of her hills. But her people remember her their 'Rani Guidallo...and a day will come when India will also remember her..." And hence, she became the 'Rani of Nagas.' He also wrote to British MP Lady Astor, requesting her to help Rani Gaidinliu, but the Secretary of State for India turned down the plea.

After the Interim Government of India was established in 1946, Rani Gaidinliu was released from Tura jail on Prime Minister Nehru's instructions following a 14-year sentence in various prisons. After her release, she continued to work for the advancement of her people. She lived in the Vimrap village of Tuensang with her younger brother Marang until 1952. In 1953, Prime Minister Nehru visited Imphal where Rani Gaidinliu conveyed her people's gratitude and goodwill to him.

During her time in Kohima, she received the "Tamrapatra Freedom Fighter Award" in 1972, the Padma Bhushan in 1982 and the Vivekananda Seva Award in 1983. Rani Gaidinliu has also been conferred the Birsa Munda Award posthumously. The Government of India issued a postal stamp in her honour in 1996. The Indian Coast Guard commissioned a Fast Patrol Vessel "ICGS Rani Gaidinliu" in 2016. On her birth anniversary in 2015, the Centre issued commemorative coins of Rs 100 and a circulation coin of Rs 5 in her honour. 'Rani Gaidinliu Tribal Freedom Fighters Museum' in Manipur's Tamenglong district, was inaugurated in November 2021. It honours her and many more unsung tribal heroes of Manipur and aims that the legacy of their struggles and sacrifices pass on to future generations.

Let us all be brave. Let us all be fighters. Let us all be Queens. Rani Gaidinliu taught us the essence of discovering the indomitable spirit to fight for what you believe in.

SANTI GHOSE
(1916–1989)

SUNITI CHOUDHURY
(1917–1988)

33-34

SANTI GHOSE AND SUNITI CHOUDHURY

Those pretty little eyes turned into fiery eyes

Santi Ghose (also known as Santi Ghosh) and Suniti Choudhury were typical teenage girls who enjoyed playing outdoors and with toys, and who may have had a pleasant and normal childhood too. But, no one expected them to shoot Charles Geoffrey Buckland Stevens, a tyrannical British magistrate, to avenge the executions of Bhagat Singh, Rajguru and Sukhdev. Santi Ghosh was only 15 years old and Suniti Choudhury 14 when they jointly shot at Stevens, killing him on the spot. The sound of gunshots alerted the British police. Soon they were overpowered and arrested. The two girls were mercilessly beaten. But they maintained their calmness all the while. They didn't panic. The girls were taken into custody and imprisoned in the local British jail, and being a minor, both of them were sentenced to jail for 10 years. They expected to die martyr's death. They stated, "it is better to die than live in a horse's stable". Both are known for their participation in an armed revolutionary struggle.

Santi Ghose was born on 22 November 1916, in Calcutta (Kolkata), to patriot Debendranath Ghose and his wife. Debendranath Ghose also taught philosophy at Victoria College in Comilla, Bengal. Ghose was raised in the Bangladeshi city of Comilla. She was raised in a patriotic atmosphere. Debendranath Ghosh, her father, was a liberation warrior and professor of philosophy.

She attended Faizunnesa Balika Vidyalaya for her studies. Ghose met Profullanandini Brahma who was the Jugantar Party's president. She was encouraged to participate in the liberation struggle not only by her family but also by Prafullanandini Brahma, who was two years her senior. Ghose was profoundly affected by Brahma. Jugantar Party was an organisation that advocated the use of violence to overthrow the British and was responsible for the murder of numerous British officials. Ghose joined the revolutionary Jugantar Party after Brahma inspired her to do so. At the age of 15, Santi Ghosh co-founded and served as Secretary of the Chhatri Sangha (Girl Students Association).

One can only imagine the level of patriotism exhibited by Brahma, 17, and Ghosh, 15, who were willing to engage in armed rebellion against the British without fear of repercussions. Santi Ghosh trained in self-defenses and the use of weapons after joining the Jugantar Party, focusing on swords, clubs, and firearms. She appeared slightly older than her age since she wore a white cotton sari to make herself appear older. Many members of the Jugantar Party were arrested by the British, handed over to the Andamans, or deported for life to Kaalapani, the Andaman Islands' Cellular Jail.

On 22 May 1917, Suniti Choudhury was born in Comilla District of West-Bengal (present-day Bangladesh) to Umacharan Choudhury and Surasundari Choudhury. She too was a student at Comilla's Faizunnesa Balika Vidyalaya. Ullaskar Dutta, who also resided in Comilla and produced bombs to be used against British colonial officials, had a profound impact on Choudhury.

From an early age onwards, Suniti resented the British. She was from a family of independence fighters. Two of her older brothers were involved in the liberation movement. She mingled with friends who shared similar beliefs on participation in India's war for independence and violent resistance against the British. She met Santi Ghose and Prafulia Nandini Brahma during her school years and was influenced by their patriotism.

Prafulla Nalini Brahma, her immediate senior at Faizunnisa Girls High School, also became a major influence on Choudhury during this period. She provided Suniti and Santi with books restricted by the British and revolutionary literature, in addition to mentoring her. The famous words of Swami Vivekanand, "Life is a sacrifice for the Motherland," inspired Choudhury's convictions.

Like Santi, Prafulla also recruited Suniti to the Jugantar Party. She also joined the Jugantar-affiliated Tripura Zilla Chhatri Sangha, the organisation's female wing. Suniti Choudhury was appointed as the District Volunteer Corps' major. When Netaji Subhash Chandra Bose was in town to address the student organisation, she led the girls in a parade. Choudhury was appointed Captain of the Women's Volunteer Corps during the Annual Conference of the Tripura

Zilla Chhatri Sangha on May 6, 1931. During this time, she was also known as "Meera Devi." She was responsible for instructing female members (of the Chhatri Sangha) in the use of the lathi, sword and dagger.

In the meantime, "Chhatri Sangha" was training young women, while the most intelligent and courageous trainees passed information, documents, weapons, ammunition, and money to the rebels. Nonetheless, Prafulla, Santi Ghose, and Suniti Choudhury wanted greater accountability and equal responsibilities like the boys. This inevitably prompted some top officials to question the purpose of girls picking up arms. Eventually, Biren Bhattacharjee, a leader who had fled from the British, had a discreet interview with the three girls. He was very impressed with their courage and encouraged them to undergo special training. Under the leadership of Akhil Chandra Nandi, president of the Tripura Students Organisation, they began their practical training. They missed school, snuck out of the congested city, and practiced shooting in the Maynamati Hills.

The most difficult aspect of shooting was managing the back kick of the revolver. Suniti's index finger was unable to reach the trigger, but she did not quit. She employed her lengthy middle finger to fire fatal shots from a tiny Belgian-made pistol. Up to this point, men were mostly responsible for the bombings and killings of British personnel. Women worked as fighters for the struggle but remained in the backdrop. Santi Ghose and Suniti Choudhury were picked for direct action to assassinate the District Magistrate of Comilla, Charles Geoffrey Buckland Stevens, on account of their gallantry and fearlessness at such a young age.

Stevens came to India on the 18th of December, 1914. On March 6, 1930, he was appointed District Magistrate and Collector of Tiepera, which was the district of Comilla. Stevens was somebody who would do anything to ruin Satyagraha. He imprisoned all the leaders and hounded every non-violent Indian who spoke up.

The freedom movement in Bengal took a serious turn at this time. Many freedom fighters protested publicly, resulting in their imprisonment. A group of patriots under the banner of a secret organisation began assassinating hateful British police officers who were known for being brutal against arrested freedom fighters. They were infamous for beating the men unconscious while in prison, subjecting them to various forms of suffering and ridiculing women Satyagrahis in the streets. Additionally, there were rumors that British police ill-treated women. Satyagrahis Santi Ghose and Suniti Choudhury were more determined than ever before to act. It was time someone stepped up and something had to be done to restore equilibrium. Santi-Suniti intended to carry it out. On March 23, 1931, news of the execution of Bhagat Singh, Rajguru, and Sukhdev shocked the nation. This violence and heartlessness on the part of the British Raj were intended to strike fear into the minds of fellow independence fighters but to no avail. One thousand Bhagat Singhs, Raigurus, and Sukhdevs sprang across the nation. As Bengal's animosity grew, Santi Ghose, Suniti Choudhury, and other young independence fighters became more willing to attack and kill British officials.

On 14 December 1931, a carriage halted in front of the bungalow of the District Magistrate. Two teenagers dismounted from it and the coachman sped off. These teenagers were Santi Ghose and Suniti Choudhury. They requested permission to speak to Charles Geoffrey Buckland Stevens, the Comilla District Magistrate. Their request was approved in response to their petition for permission to organise a swimming competition among their peers at school. They entered the office where the magistrate was seated. They concealed their automatic weapons beneath their shawls. The British authorities had no reason to suspect them and refrained from searching them as they were simple school girls. This made their entry easy. Soon they were ushered inside. Stevens scanned the letter that was handed to him. According to their signatures, Illa Sen and Meera Devi petitioned the Magistrate for a swimming club. The use of "Your Majesty" and other instances of poor English reflected their sincerity and simplicity. The girls were growing eager for a window to be cleaned. They requested Stevens' signature on the letter as a recommendation. He left his office and returned shortly with the signed document. That was his final action before guns rang out in the house. The legendary judge's final vision was of two young girls, now without their shawls, pointing two revolvers directly at his heart. Santi Ghose and Suniti Choudhury fired at Stevens, instantly killing him. The SDO intervened too late. The British police were alerted by the sound of gunfire. They were quickly subdued and arrested. When police arrived to apprehend the two adolescents, Santi and Suniti did not resist the severe beating that happened. They had started on this mission prepared for the forthcoming torture by pricking their fingertips with pins to strengthen their pain tolerance, so they would not succumb to beatings or reveal the identity of their secret group. Even when they were assaulted under the guise of weapon searches, they remained composed.

They anticipated a martyr's death. They sang patriotic songs from the police vehicle to the courtroom and back again, delivering beautiful smiles and blessings to those gathered nearby and far away. Their smiles only faded when the verdict was announced. It was life imprisonment. Due to their young age, they were spared the executioner's noose. The girls were upset that they did not receive the death penalty since they desired to be martyrs for the cause of freedom.

The authorities never relented in their efforts to subdue them. Prafulla was arrested on suspicion of being the primary conspirator. Later, she was placed under rigorous house arrest, where she died from a lack of medical treatment after five years. The first bullet fired from Suniti Choudhury's revolver instantaneously caused the death of the British officer. The next volley of gunshots fired by the two teenage females rendered the Magistrate's survival impossible. As a result, Suniti received a heavier sentence and was isolated from the rest of the political prisoners.

Suniti was sent to third class with the criminals and pickpockets, while Santi remained in second class with the other revolutionaries. Suniti's family was also affected by her actions since her father's government pension was terminated and her two eldest brothers were detained without trial. Years of hunger contributed to her younger brother's death from consumption, which was compounded by malnutrition.

They were released in 1939 under an amnesty agreement between the government and Mahatma Gandhi. But the revolutionaries never quit. Ghose joined the Indian National Congress following her release from prison. She served on the West Bengal Legislative Council from 1952 to 1962 and again from 1967 to 1968, as well as the West Bengal Legislative Assembly from 1962 to 1964. She wrote a book titled Arun Bahni. She participated in communist movements as well.

After her release, Choudhury worked hard and passed the M.B.B.S. examination. In 1944, she gained admission to Calcutta Medical College after completing the Licentiate in Medicine and Surgery (LMS) at Campbell Medical School. Then she wed Pradyot Kumar Ghose, an activist and former political prisoner who was also a friend of her brother. In Chandannagar, where she was affectionately referred to as "Lady Maa, "Suniti's skill, combined with her compassion and dedication, swiftly made her a well-respected doctor (Lady Mother).

Santi took her last breath in 1989 and Suniti breathed her last in 1988, leaving behind a legacy of extraordinary valor, selflessness and nationalism.

You need to have that highest level of passion where you are not willing to let bygones be bygones. When you know you are not immune to life, but have your own thoughts and feeling about the Motherland and want to make it free and safe for all of us. That's what we learn from these two girls who had sacrificed their childhood and their life for the nation's independence.

KANAKLATA BARUA
(1924–1942)

35

KANAKLATA BARUA

The Valiant Martyr

How many of us want to fight to make the world a better place? At the age of only 17 years, Kanaklata Barua from Assam knew where her heart was and wanted to get things done to free her country and give her people a better life. Patriotic fervor was building up in the country and she didn't want to be left behind. Kanaklata joined a suicide squad called Mrityu Bahini consisting of young volunteers. On September 20, 1942, the revolutionary camp of the Gohpur division of undivided Darrang district, led by Jyoti Prasad Agarwala, planned to raise the National Flag at the local police station in commemoration of Kushal Konwar's martyrdom. Kanaklata marched in front, leading a throng of female volunteers. She approached the police station premises fearlessly. The police opened fire without warning and Kanaklata was shot in the chest. She proudly held the Tricolor up with her blood-stained young hands as she fell to the ground.

Kanaklata Barua was born on December 22, 1924, in Barangabari village of Gohpur sub-division of Sonitpur district to Krishna Kanta Barua and Korneshwari Barua. Her mother passed away when she was just five years old and the little girl was suddenly loaded with responsibility. Her father remarried after her mother died. Unfortunately, her father died as well when she was thirteen. With household responsibilities and siblings to look after, Kanaklata had no choice but to drop out of school.

The spirit of the freedom struggle expanded throughout the country following the horrible massacre at Jallianwala Bagh in 1919. In 1921, the freedom movement took a new direction. Throughout the country, a non-cooperation movement was launched to encourage the British to give Swaraj. This national uproar spread like wildfire, sparking patriotism in the hearts of the people. The movement resulted in a boycott of all British social, political, economic and educational institutions. Following the Chauri Chaura event, the movement was disbanded.

In 1942, the liberation movement's activities reached a zenith, with Gandhi launching the "Quit India Movement" with the rallying cry "Do or Die." Assam too felt the patriotic waves of this national upsurge, and there were numerous protests against British control. India's struggle for independence from colonial rule was drawing to a close, and women from all over the country were playing a critical role in it. These women hugely inspired this 17-year-old and she was all in for her country.

A big camp of revolutionaries was established in Assam's Tezpur district, some 8 km from Kanaklata's home. Comrade Bishnu Prasad's speeches and Jyoti Prasad Agarwala's songs, a well-known Assamese cultural icon, had a strong impact on her patriotic impulses. This was her moment and Kanaklata began secretly attending camp sessions. Now there was no turning back.

Kanaklata's grandfather having been through the painful loss of his son and daughters-in-law, knew what revolutionary patriotism required sacrifice, and rejected her decision to join revolutionaries. He tried to reason with her, but Kanaklata had a mind of her own. She was adamant and persistent; she convinced her stepmother who allowed her to join the revolutionaries officially. Soon, her younger brother, Rajanikanta, too volunteered with the Quit India Movement.

Kanaklata joined Shanti Bahini (peace force) consisting of young volunteers, who were tasked with guarding the villages at night. Though movements were relatively peaceful in Assam, the mood was volatile. Jyotiprasad Agarwala, a writer, musician, and playwright, was a prominent figure in Assam's freedom struggle. He urged Mukti Bahini members to peacefully march to Tezpur district police stations to hoist the Indian Flag and raise the "Quit India" slogan. As a result, he established the Mrityu Bahini for this objective, which adhered to Gandhi's ideology of "do or die." Kanaklata was a member of this brigade that was aware of the fact that death was inevitable. Mrityu Bahini was a formidable force. They were all inspired by contemporary Assamese freedom fighters such as Kushal Konwar, the then-President of the local Congress Committee who was falsely accused of the train derailment by the British and hanged.

On September 20, 1942, the revolutionary wing of the Gohpur sub-division of the undivided Darrang district resolved to remove the British Flag from the local police station in Gohpur and replace it with the Indian national Flag. Kanaklata was to lead a parade of 5,000 unarmed villagers and youngsters, most of whom were students, in this tough task. Around 500 females from the surrounding villages joined the march.

Winding up her routine work, she fulfilled her home duties and shared a meal with her younger brother and sisters. She sorrowfully informed them that she might not come back alive. Kanaklata vowed to her grandfather that she would uphold the proud reputation of his Ahom ancestors as she led the protesters towards the police station to unfurl the Indian Flag, and then she went out to join the protesters. The entire route to the police station was filled with fervor as Vande Mataram slogans were called out and instruments like conch shells, drums, flutes, etc., were played. Rebati Mahan Som, the officer in command of the police station, had anticipated the protester's actions. His armed men were ready to fire. He continuously warned the throng of the dire consequences that would follow if they carried out their plan. Kanaklata sharply instructed him to perform his duties while she continued with hers. Her impassioned address, delivered just hours before her death, had captivated the audience. As soon as they entered the police station, without any warning the police opened fire. Kanaklata was shot in the chest while holding the Flag and ensuring that it did not fall to the ground until it was taken up by her comrade Mukunda Kakoti, who was also slain in the gunfire.

Hemakanta Baruah, a sixteen-year-old minor school boy, met the same end. Several other volunteers sustained significant injuries.

Both Kanaklata and Mukunda's efforts, sacrifice and bravery were not in vain, as Rampati Rajkhowa eventually unfurled the Tricolor at the police station in the evening, bringing extra support for the independence struggle and achieving what his fellow protestors had given up their lives for. Mrityu Bahini's brave move bolstered India's will to overthrow and undermine England's control of the country, till independence was declared on August 15, 1947.

Kanaklata Barua's heartbreaking martyrdom at a tender age established her as a household name in Assam. She has posthumously bestowed the titles of 'Shaheed' and 'Birbala', meaning 'martyr' and 'brave woman', respectively, the first of numerous honours. Even today in Assam, the name 'Kanaklata' symbolizes revolution against oppression and injustice. Her memory lives on in Assam's sociopolitical and cultural spheres to this day. As a tribute to her valor during the Freedom Struggle, the Indian Coast Guard's Fast Patrol Vessel ICGS, commissioned in 1997 was named after Kanaklata Barua. It engaged in several search and rescue missions, the capture of foreign fishing vessels intruding into Indian waters, and rescue missions during its two decades of duty. Her statue was installed in 2011 in 'Kanaklata Udyan' in Gauripur (formerly Gohpur), Tejpur. The teenage patriot's extraordinary bravery has also been immortalised in the Assamese film Epaah Phulil Epaah Xoril, directed by Chandra Mudoi, and released in Hindi as Purab ki Aawaz.

The rebelliousness of this teenage brave-heart was frowned upon in those days, but her enormous courage and ultimate sacrifice have left an indelible impression on Assam and the entire country, inspiring people of all ages.

To become inspiration to others, you need to portray pure courage as Barua displayed in her teens. She developed a sense of responsibility at the tender age of five when her mother passed away. The valiant act of Mrityu Bahini only strengthened India's resolve to dethrone and further weaken England's grip is a lesson worth learning.

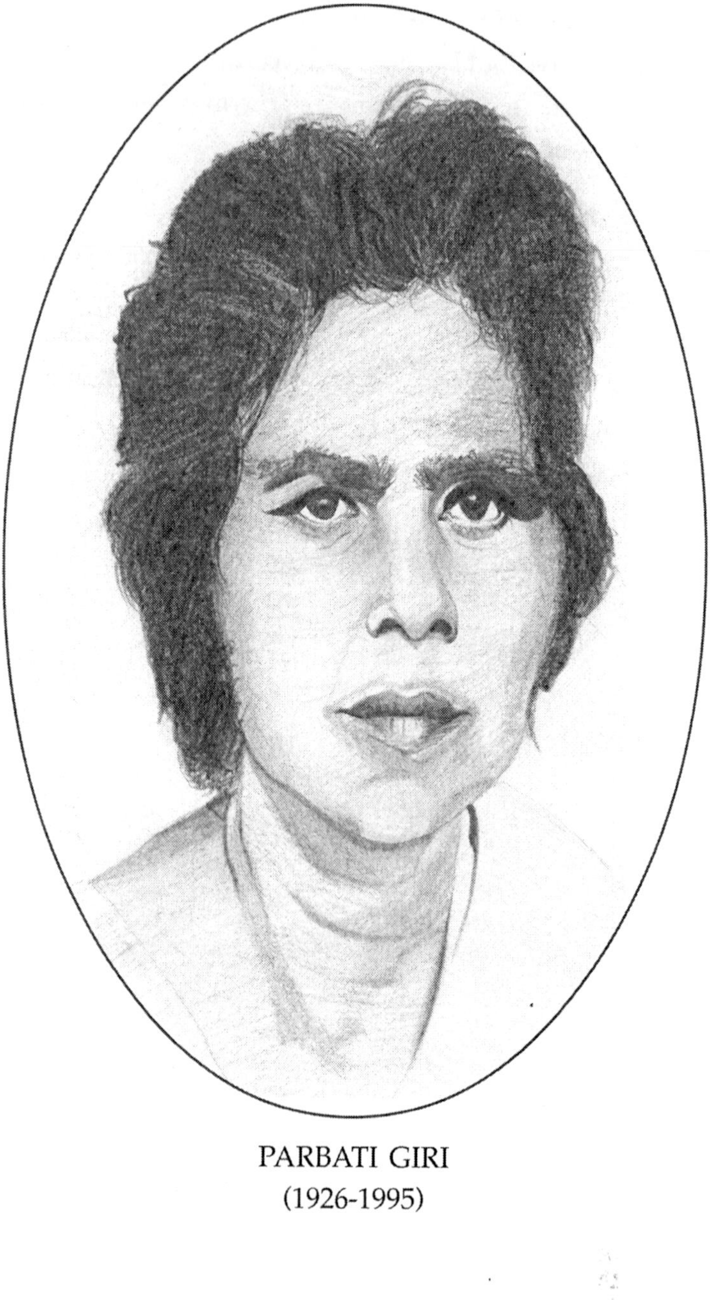

PARBATI GIRI
(1926-1995)

PARBATI GIRI
(1926-1995)

36

PARBATI GIRI

A fierce heart that doesn't settle within the four walls

Parbati Giri was a notable woman who played a heroic part in the Indian Independence Movement. Due to her anti-government actions, she was arrested and sentenced to two years in prison. The legendary woman from the Western Odisha area of Bargarh is affectionately called Badamaa (Big mother) by the residents of her Ashrams. Parbati was also called "Banhi Kanya" (daughter of fire) for her persistence, fierce nationalistic emotions and freedom of expression. She laboured for the underprivileged till her death.

Parbati Giri was born on January 19, 1926, at Samlaipadar village in the Bargarh district of Odisha. Her parents were Srimati and Dhananjay Giri. She was filled with patriotism since childhood. Samlaipadar village was the epicenter of political activity at that time when nationalist feelings were sweeping over Odisha. Parbati Giri's uncle Ramachandra Giri was among the prominent leaders who were meeting and devising methods to oppose the British establishment. Parbati frequently accompanied her uncle to these sessions, where she was profoundly influenced by the concepts and ideas being discussed. These sessions influenced her choice to join the nation's independence movement.

Although Parbati desired to pursue further education, she dropped out of school at age 11 and began working for the local assembly of congress workers. She travelled from village to village as a child also raised awareness and campaigned for Congress. In the year 1937, a Congress conference was held in Samlaipadar. A large gathering of Congress workers was convened at Samlaipadar just one year later, in 1938. Distinguished figures such as Malati Choudhury and Pranakrushna Padhiari delivered important addresses. Senior leaders of the Congress appealed to Parbati's father to permit her to participate in the mission, citing her dedication and remarkable work. Parbati Giri was allowed to visit Rama Devi's Bari Ashram after passing a tough test.

Parbati Giri left home on 14th January 1938. A girl's journey from Samlaipadar, a small and inconsequential village, to Jajpur's Bari Ashram was a bold one. Prabhabati Devi, another child widow from Bargarh, volunteered to accompany Parbati Giri at that time. Prabhabati's family members also attempted to encourage both daughters to return home. However, their efforts were in vain. Despite these challenges, they continued to Bari Ashram. On the route, they stopped at Reba Roy's house in Arigaon. The following day, they arrived at the Bari Ashram. They both met Rama Devi, Gopabandhu Choudhury, and other individuals. The ancient Ashrams were founded on a system of strong moral values, which cannot be compared to modern

Ashrams. The names of these Ashrams were "Sangrami Sibira" and "Ahimsa tirtha". Gopabandhu Choudhury was also a remarkable man, he resigned from his position as Deputy Collector and devoted himself to the liberation cause.

Thus, Parbati Giri was strongly impacted by the beliefs of Gopabandhu Choudhury and Rama Devi, as well as Mahatma Gandhi. She was instructed in weaving, crafting, and the significance of social service. She accepted Gandhism as a lifestyle, and her work reflected this quality. When direct Satyagraha began in 1940, she began organising meetings and mobilising villagers to engage in Gandhi's Khadi Andolan movement. She persuaded them to pursue charkha spinning and wear khadi. In 1942, the Quit India Movement was initiated. Parbati was just 16 when she was at the forefront of agitation following Gandhi's 'Quit India' movement. Very young Parbati Giri joined the other leaders of the movement. She participated in protests waving the Tricolor Flag and chanting anti-British slogans, never hesitating to demonstrate her staunch hostility toward the British elite.

She had also staged an agitation at Bargarh Court to persuade the lawyers to boycott the erstwhile court in defiance of the British. She reportedly even went to the Bargarh Court and told the attorneys to leave. Her objective was to convince them to cease working with British officials on legal concerns. Those who disregarded her instructions were given two bangles as a mark of their cowardice. She and three other boys' arrival in the SDO's office in Bargarh was another remarkable occurrence. She sat on the SDO's chair while pretending to be a judge. As soon as she saw the SDO enter the office, she instructed the other boys to bring him to inbound like a criminal. Due to this absurd and brazen act, she was arrested and sentenced to two years in prison.

Parbati Giri dedicated her life to humanitarian work after India's independence on August 15, 1947. She founded an Ashram for the underprivileged and oppressed. She worked extensively to improve prison conditions and eradicate leprosy. She also took part in Binoba Bhave's Bhudaan movement. Parbati Giri was offered an assembly ticket and a Rajya Sabha seat, but she politely declined and dedicated her life to the benefit of the impoverished and oppressed.

Her fearless attitude toward the British was directly opposed to the kindness and determination with which she laboured for people's well-being. When Odisha was devastated by hunger in 1951, she took up the task of relief efforts and assisted the most vulnerable. Her Ashram door was open for the leprosy patients who are still looked down upon and live an isolated life.

Parbati established the Rukmini Lath Balniketan for orphans and children in Paikmal village, Bargaon. She also founded the Kasturba Gandhi Matru Niketan in Paikmal, Bargarh district, which is still in operation and serves the impoverished. Despite having contacts with then-political elites, she never took advantage of them.

In 1984, the Department of Social Welfare of the Government of India gave Parbati a prize for her dedicated social service. In 1988, she was also awarded an honourary doctorate by the University of Sambalpur. The Odisha government has also named an irrigation project after her, the 'Parbati Giri Mega Lift Irrigation Scheme.' The great Parbati Giri died on August 17, 1995, after a prolonged illness.

From being a school drop-out, Parbati Giri set an example of an energetic teen who joined the Quit India Movement at a young age of 16. From her life it is evident that selfless contribution during freedom struggle, and a combination of fearless attitude and kindness are characteristics which made her a woman with strong heart and determined mind.

WEBSITE REFERENCES

https://feminisminindia.com/2017/03/13/rani-gaidinliu-essay/, https://theprint.in/features/rani-gaidinliu-daughter-of-the-hills-who-spent-14-years-in-jail-for-indias-independence/591570,

https://feminisminindia.com/2017/06/13/kittur-chennamma-essay/

https://feminisminindia.com/2018/03/30/bhikaji-cama-exile-indian-flag/

https://feminisminindia.com/2018/07/31/kuyili-dalit-commander-sivagangai/

https://feminisminindia.com/2018/10/17/velu-nachiyar-queen-freedom-struggle/

https://feminisminindia.com/2019/09/10/suniti-choudhury-youngest-female-revolutionary-india/

https://indianculture.gov.in/node/2790450

https://indianculture.gov.in/node/2794858

https://indianexpress.com/article/cities/delhi/freedom-fighter-satyavati-mother-of-exvp-dies-at-105/

https://indianexpress.com/article/explained/icgs-kanaklata-barua-indian-coast-guard-6654728/

https://indianexpress.com/article/explained/rani-gaidinliu-tribal-freedom-fighters-museum-manipur-7637274/,

https://indianexpress.com/article/lifestyle/art-and-culture/bina-das-21-yr-old-who-shot-bengal-governor-got-padmasri-but-died-in-penury-6298713/

https://indianexpress.com/article/lifestyle/art-and-culture/pritilata-waddedarrevolutionary-india-the-21-year-old-who-chose-to-die-than-be-caught-by-the-british-6300316/

https://indianexpress.com/article/parenting/learning/rani-lakshmibai-death anniversary-2019-why-we-remember-jhansi-ki-rani-5786199/

https://indianliberals.in/content/ramadevi-choudhury/

https://indianliberals.in/content/ramadevi-choudhury/
https://www.google.co.in/books/edition/NATIONALIST_MOVEMENT_IN_ODISHA/HCmjDwAAQBAJ?hl=en&gbpv=1& amp;dq=Prabhavati+devi&pg=PA29&printsec=frontcover
https://kalingatv.com/features/remembering-freedom-fighter-parbati-giri-of-odisha/
https://lifebeyondnumbers.com/17-women-freedom-fighters-of-india-who-epitomized-fearlessness/
https://magazines.odisha.gov.in/Orissareview/aug2005/engpdf/parbati%20giri.pdf
https://satyaagrah.com/sanatan/veer-gaatha/785-santi-ghosh-and-suniti-choudhury-two-teenage-freedom-fighters-assassinated-british-magistrate
https://scroll.in/article/1020396/ramachandra-guha-what-the-rss-could-learn-about-nationalism-from-freedom-fighter-usha-mehta
https://starsunfolded.com/bhikaiji-cama/
https://thelogicalindian.com/history/velu-nachiyar-tamil-nadu-33540
https://theprint.in/opinion/the-political-dynasty-nobody-is-talking-about-sardar-patels/142422/
https://theprint.in/pageturner/excerpt/freda-bedi-the-english-girl-married-to-a-punjabi-who-joined-indias-freedom-struggle/214143/
https://theprint.in/theprint-essential/remembering-capt-lakshmi-freedom-fighter-social-activist-on-her-6th-death-anniversary/86907/
https://theprint.in/theprint-profile/aruna-asaf-ali-a-fiercely-independent-freedom-fighter-who-defied-mahatma-gandhi/90230/
https://theprint.in/theprint-profile/kasturba-gandhi-the-feisty-woman-whose-patience-inspired-gandhis-call-for-satyagraha/218949/
https://wbchse.nic.in/html/matangini_hazra.html
https://www.amazon.in/Stories-Must-Tell-Emotional-Actor/dp/9390679400
https://www.ambedkaritetoday.com/2018/11/about-jhalkari-bai-biography-and-life-history-of-dalit-woman-warrior.html
https://www.ambedkaritetoday.com/2019/01/uda-devi-a-dalit-women-freedom-fighter.html Uda Devi: Dalit Freedom Fighter

In The 1857 Uprising | #IndianWomenInHistory
https://www.amritmahotsav.nic.in/unsung-heroes-detail.htm?40
https://www.bbc.co.uk/history/historic_figures/besant_annie.shtml
https://www.bbc.com/news/world-asia-india-46943612
https://www.bbc.com/news/world-asia-india-53775313
https://www.britannica.com/biography/Annie-Besant,
https://www.britannica.com/biography/Bhikaiji-Cama
https://www.chhatrashakti.in/2020/08/04/unsungfreedomfighter-tara-rani-srivastava/
https://www.constitutionofindia.net/blogs/assembly_member_of_the_week___sucheta_kripalani
https://www.constitutionofindia.net/constituent_assembly_members/vijaya_lakshmi_pandit
https://www.culturalindia.net/leaders/sarojini-naidu.html
https://www.deccanherald.com/opinion/in-perspective/kamaladevi-how-could-we-forget-908413.html
https://www.digpu.com/featured-posts/story-of-matangini-hazra-gandhi-buri
https://www.encyclopedia.com/people/history/south-asian-history-biographies/vijaya-lakshmi-pandit
http://magazines.odisha.gov.in/Orissareview/April2006/engpdf/freedom_struggle_%20and%20rama%20devi_.pdf
http://www.iloveindia.com/indian-heroes/begum-hazrat-mahal.html
http://www.kamat.com/kalranga/women/kundapur.htm
https://aidindia.org/bibi-gulab-kaur/
https://alchetron.com/Basanti-Devi
https://biography.yourdictionary.com/vijaya-lakshmi-pandit
https://byjus.com/free-ias-prep/aruna-asaf-ali-1909-1996/
https://byjus.com/free-ias-prep/this-day-in-history-nov22/
https://cmsadmin.amritmahotsav.nic.in/unsung-heroes-detail.htm?278
https://cmsadmin.amritmahotsav.nic.in/unsung-heroes-detail.htm?96
https://www.fredabedi.com/excerpt.html
https://www.google.co.in/books/edition/NATIONALIST_MOVEMENT_IN_ODISHA/HCmjDwAAQBAJ?hl=e n& gbpv =1&dq=Prabhavati+devi&pg=PA29& printsec=frontcover

https://www.hindujagruti.org/articles/51_kittur-rani-chennamma.html

https://www.hindustantimes.com/trending/bina-das-unsung-and-unforgettable-101629793944600.html

https://www.iloveindia.com/indian-heroes/kamala-nehru.html

https://www.indianculture.gov.in/stories/begum-hazrat-mahal-revolutionary-queen-awadh

https://www.indianetzone.com/2/basanti_debi_chittaranjan_das.htm

https://www.indiatoday.in/education-today/gk-current-affairs/story/know-all-about-rani-lakshmibai-of-jhansi-born-as-manikarnika-tambe-1316804-2018-08-17

https://www.indiatoday.in/education-today/gk-current-affairs/story/remembering-nehru-s-sister-vijaya-lakshmi-pandit-who-was-imprisoned-thrice-during-the-indian-freedom-struggle-1317554-2018-08-18

https://www.indiatoday.in/education-today/gk-current-affairs/story/kamala-nehru-119th-birth-anniversary-1302267-2018-08-01

https://www.indiatoday.in/education-today/gk-current-affairs/story/parbati-giri-the-mother-teresa-of-odisha-left-school-in-class-3-to-fight-for-india-s-freedom-1317752-2018-08-18

https://www.indiatvnews.com/news/india/who-was-bhikaji-rustom-cama-story-flag-of-indian-independence-foreign-soil-freedom-struggle-641704

https://www.inuth.com/india/kamala-nehru-how-jawaharlal-nehrus-shy-wife-fought-against-the-british/

https://www.jagranjosh.com/general-knowledge/sarojini-naidu-biography-1646201899-1

https://www.keralapscgk.com/2020/01/famous-personalities-bhikaiji-rustom.html

https://www.livehistoryindia.com/story/herstory/begum-hazrat-mahal-a-revolutionary-queen/

https://www.livehistoryindia.com/story/herstory/velu-nachiyar/

https://www.livehistoryindia.com/story/history-daily/kanaklatabarua/, https://feminisminindia.com/2019/03/14/kanaklata-barua-teen-freedom-fighter/, http://www.assaminfo.com/famous-people/32/kanaklata-barua.htm,

https://www.livehistoryindia.com/story/people/umabai-kundapur-an-unsung-hero

https://www.liveindia.com/freedomfighters/jhansi_ki_rani_laxmi_bai.html

https://www.mapsofindia.com/who-is-who/history/aruna-asaf-ali.html

https://www.organiser.org/india-news/remembering-birbala-kanaklata-barua-on-her-79th-death-anniversary-5581.html

https://www.shethepeople.tv/sepia-stories/meet-durga-devi-woman-who-played-vital-role-in-indias-fight-for-freedom/

https://www.shethepeople.tv/sepia-stories/meet-durga-devi-woman-who-played-vital-role-in-indias-fight-for-freedom/

https://www.thebetterindia.com ›Stories

https://www.thebetterindia.com/11547/doctor-freedom-fighter-activist-feminist-inspiring-womalakshmi-sehgal/

https://www.thebetterindia.com/138291/india-first-woman-cm-freedom-fighter-sucheta-kriplani/

https://www.thebetterindia.com/155824/news-west-bengal-pritilata-waddedar-kalpana-datta-chittagong/

https://www.thebetterindia.com/190053/gulab-kaur-woman-freedom-fighter-punjab-independence-day-india/

https://www.thebetterindia.com/69290/madam-bhikaji-cama-flag-stuttgart-india/

https://www.thebetterindia.com/94158/kamaladevi-chatto padhyay-feminist-freedom-fighter-cooperatives-faridabad/

https://www.thebetterindia.com/97013/jhalkari-bai-laxmibai-jhansi-india-freedom/

https://www.thefamouspeople.com/profiles/aruna-asaf-ali-5422.php

https://www.thefamouspeople.com/profiles/aruna-asaf-ali-5422.php

https://www.thefamouspeople.com/profiles/begum-hazrat-mahal-5518.php

https://www.thefamouspeople.com/profiles/kasturba-gandhi-5481.php

https://www.thehindu.com/society/history-and-culture/united-colours-of-freda-bedi/article26280964.ece

https://www.udayavani.com/english-news/kuyili-a-brave-commander-who-was-the-first-ever-suicide-bomber-in-history

https://www.varanasi.org.in/annie-besant

https://www.womensweb.in/articles/inspiring-woman-aruna-asaf-ali/

INDEX

Abhinav Bharat, 47
Achyutrao Patwardhan, 115
Ahimsa Tirtha, 201
Ahimsa, 54
Ali, Aruna Asaf, 141-42, 144-46
Ali, Hyder, 26
All India Trade Union Congress, 146
Allahabad Municipal Board, 108
All-India Congress Committee, 79, 125, 144
All-India Democratic Women's Association (AIDWA), 173
All-India Mahila Congress, 126
All-India Minorities Front, 146
America: The Land of Superlatives, 103
Ammukutty, A.V., 171
Andaman, Kala Pani Jail, 48
Ashafaqullah, 149
Asprushyata Nibarna Samiti, 85
Avabai Petit Orphanage, 51
Awadh, 12-16, 29
Azad, Chandra Shekhar, 120, 149

Baba Hafiz Abdullah (Fajja), 63
Baba Harnam Singh (Tundilat), 63
Babu, Phulendu, 131
Badamaa, 199
Bai, Jhalkari, 32-36
Bailhongal Fort, 9, 10
Bande Mataram, 49
Banhi Kanya, 199
Bardoli Swaraj Ashram, 96
Bedi, Baba Pyare Lal, 163
Bedi, Pyare Lal, 164
Bengal Provincial Congress, 74
Bengal Provincial Legislative Assembly, 159
Besant, Annie, 37-39, 41-42, 67, 99
Bhagini Samaj, 79
Bhagwad Gita, 120
Bharatiya Natya Sangha, 104
Bhaumik, Jatish Chandra, 160
Bhave, Vinoba, 84
Bhave's Sarvodaya movement, 86
BHU, 125
Bombay Parsi Community, 50
Bose, Subhash Chandra, 74-75, 157, 170, 172
Brahmo Samaj, 143
British, 4, 9, 21, 26, 35, 80, 124, 139, 144
British Army, 26-27
British East India Company, 7-8, 12, 20, 29
British Flag, 195
British Raj, 112, 187
Bundelkhand Folklore, 35
Burma, 167

C. Rajagopalachari, 104
Cama, Bhikaji Rustom, 44-45, 49, 51

Campbell's 93rd Highland Regiment, 30
Cape Supreme Court, 53
Captain MacDonald, 178
Caribbean Island, 50
Central Cottage Industries Emporium, 104
Central Hindu College in Benares, 41
Chand, Babu Purna, 15
Char Bagh, 16
Charles Geoffrey Buckland Stevens, 188
Charles Knowlton's Fruits of Philosophy, 38
Chhatri Sangha, 156-57, 186
Chicago Radio, 115
Chitaranjan Das, 73
Chittagong Armory Raid, 137-38
Choudhury, Gopabandhu, 83-84, 201
Choudhury, Rama Devi, 82-83
Choudhury, Suniti, 185-90
Christabel Pankhurst, 49
Civil Disobedience Movement, 59-60, 69-70, 73
Colonel Prem Kumar Sahgal, 173
Communist Party of India (Marxist), 145, 170, 173
Congress Socialist Party, 146
Craft Council of India, 104
Criminal Investigation Department (CID), 116, 151

Dalit Veeranganas, 29
Dandi March, 70, 96
Das, Bina, 155—59
Das, C R, 73
Das, Gopal Ballav, 84
Daughter of the Hills, 179
Dayal, Lala Har, 47
Delhi Craft Council, 104
Denmark, 102
Dera Baba Nanak, 164-65
Desai, Mallasarja, 7
Desai, Morarji, 116
Devi, Ambalika, 143
Devi, Basant Kumari, 84
Devi, Basanti, 72-75
Devi, Devki, 119
Devi, Jamuna, 33
Devi, Malati, 85
Devi, Meera, 186
Devi, Prabhavati, 132-34
Devi, Rama, 85-87
Dhanna Bhagat, 104
Dilkusha Palace, 16
Dindigul, 3
Doctrine of Lapse, 7-8, 14, 20
Dr Aruna Asaf Ali Sadbhawan Award, 146
Dr Chattopadhyaya, 68
Durgabai, 91
Durgawati Devi, 148-54

Edinburgh University, 67
Egypt, 49

Faizunnesa Balika Vidyalaya, 184
Fast Patrol Vessel *ICGS Rani Gaidinliu*, 180
Field Marshal Hugh Henry Rose, 34
First War of Indian Independence, 21
Folk Ballads (Lavani), 10
Freda Bedi, 161-62, 167

Freedom Salt, 102

Gandhi Buri, 58
Gandhi Smarak Nidhi, 117
Gandhi, Kasturba, 52-56, 85, 95, 133
Gandhi, Mahatma, 42, 53, 55-56, 69-70, 78-79, 81, 84, 102, 104, 107-08, 130, 165
Gandhi-Irwin Pact, 85, 144
Ganguly, Upendranath, 143
Gaundevi Mahila Samaj in Mumbai, 78
Gaur, Vijay, 10
Gauri Puja, 34
Gelongma Karma Kechog Palmo, 167
General Colin Campbell, 30
General Rose, 35
General Sir Hugh Rose, 22
Giri, Parbati, 198-202
Gokhale Memorial School, Calcutta, 143
Gokhale, Gopal Krishna, 69, 99
Gosse, Edmund, 68
Gowalia Tank Ground, 114, 145
Grama Sevikas, 81
Gujarat Vidyapith, 96
Gwalior, 22

Haipou Jadonang, 177
Hardikar, Dr N S, 79
Harijan Sewa Sangh, 85
Hazra, Matangini, 57-59, 61
Hazra, Trilochan, 59
Hazrat Mahal, 14
Hazrat, Munshi Ghulam, 15
Hazratganj, Old Victoria Park, 17
Helena Blavatsky, 41
Heroine of the 1942 Agitation, 145
Himalayan Toilets, 152
Hindustan Socialist Republican Association (HSRA), 149-50, 152
Hindustani Seva Dal, 79
Home Rule League, 41
Houlston, Freda Marie, 162
Hyde Park, 46
Hyderabad Ladies' War Relief Association, 68

Iftakkar-un-Nissa, 13
India House, 47
Indian Coast Guard's Fast Patrol Vessel ICGS, 196
Indian Cooperative Union, 103
Indian Flag, 45, 58, 102, 195
Indian National Army (INA), 170, 173
Indian National Congress, 38, 42, 60, 74, 84, 90, 96, 108, 143
Indian National Theatre (INT), 98
Indian Nationalist Movement, 45, 69-70
Indira Priyadarshini, 90
Inquilab Magazine, 145
International Puppetry Organisation, 105
International Socialist Conference, 45
International Socialist Congress, 47
Iyer, C.P. Ramaswami, 69

J. Jayalalitha, 4
Jackson, Stanley, 159
Jadonang, 177
Jain, Vimal Prasad, 152
Jallianwala Bagh Massacre, 77, 124

Jhansi Army, 22
Jhaveri, Chandrakant Babubhai, 114-15
Jhaveri, Vithalbhai, 115
Jhaveri, Vithaldas K, 114
Jinnah, Muhammad Ali, 69
Joan of Arc, 23
John P. Saunders Murder, 149-50
Jugantar Party, 184
Jungabai, 77

Kaiser-i-Hind medal, 70
Kakati, 7
Kalaiyar Koil War, 3
Kalam, A.P.J. Abdul, 173
Kamala Nehru Hospital, 92
Kamala Nehru Park, 92
Kamaladevi Chattopadhyay, 80, 91, 97-105
Kanaklata Barua, 191-92, 196
Kanaklata Udyan, 196
Kant, Krishan, 121
Kapadia, Gokuladas, 53
Karnataka Press, 78
Kasturba Trust's Karnataka, 81, 86
Kathmandu's Jama Masjid, 17
Kaur, Bibi Gulab, 62-65
Kesar Bagh, 16
Khadi, 86
Khan, Mohammad Ibrahim, 15
Khan, Mummu, 15
Khawasin, 13
Khilafat
  Movement, 73
  Problem, 70
Khursheed, 104
Kisan Mazdoor Praja Party (KMPP), 126
Kittur Rani Chennamma Memorial Committee, 10
Kittur Utsava, 10
Kitturu Chennamma, 10
Koli, 35
Kolkata Loreto House, 73
Kori, 33
Kriplani, Jivatram Bhagwandas (Acharya), 81, 124-25
Kriplani, Sucheta, 122-24
Krishna Nehru Hutheesing, 91
Kudanchavadi, 25
Kuyili, 3, 24-27

Labour Club, 163
Labour Movement, 39
Lahore Female Jail, 166
Lahore Fort, 63-64
Lahore Women's Jail, 166
Licentiate in Medicine and Surgery, 190
Lohia, Dr Ram Manohar, 115
Lohia, Ram Manohar, 102, 145
Lok Sabha, 96, 109, 127
Lokmanya Bal Gangadhar Tilak, 41
Lord Elphinstone, 8
Lord Hailey, 153

Mahak Pari, 13
Mahal, Begum Hazrat, 11-16
Maharaja Balakrishna, 15
Mahila Charkha Samiti, 133
Majumdar, Dinesh, 158
Majumdar, Surendranath, 123
Makka Pasi, 29
Malthusian League, 40
Maniben, 94, 95
Manikarnika Ghat, 19

Manila, Ghadar Party, 63-64
Manu, 19
Maratha Empire, 19
Marryat, Frederick, 39
Martinique, 50
Maruthu Pandiyars, 26
Master Da, 136-38
Maulana Azad, 85
Mazgaon, Framji Nusserwanjee Patel Agiary, 51
Mehta, Usha, 111, 112
Montagu-Chelmsford Reforms, 70
Moosabagh, 16
Morepant Tambe, 19
Morris, Emily, 39
Muhammadi Khanum, 12
Mumbai's Chowpatty Beach, 78
Munchershah Burjorji Godrej, 48
Muslim League, 42

Nag, Leela, 137
Naga Community, 177-78
Naidu, Padmaja
 The Feather of the Dawn, 68
Naidu, Sarojini, 66-70
 The Lady of the Lake, 68
Nanka Motwani, 115
Naoroji, Dadabhai, 46
Naradha Gana Sabha, 4
Narayan, Jayaprakash, 84, 102, 133, 145
Narayan, R.K., 174
Narayanan, Edatata, 145
Nari Karma Mandir, 73-74
Nariman Printer, 115
National Flag, 130-31, 192
National School of Drama, 104
Native Horse Artillery, 8
Navajivan Publications, 96
Navjivan Trust, 96
Nayaker, Gopala, 3
Necessary Industrialisation, 146
Nehru, Jawaharlal, 69, 85, 89-91, 107, 179
Nehru, Kamala, 89-91
Nehru, Motilal, 104
Nepal, 12
New York City, 41, 68
Nightingale of India, 67
Non-Cooperation Movement, 90, 94-95, 100, 177
North Cachar Hills, 178

Overbearing Englishwoman, 158
OVM Dance Academy, 4
Oxford Freda, 163
Oxford Majlis, 163

Pahartali European Club, 136, 138
Pall Mall Gazette, The Secret Doctrine, 41
Panajkar, Krishnabai, 79
Pandit, Vijaya Lakshmi
 So I Became a Minister
 Prison Days, 110
Panthulu, B.R., 10
Pari Khana, 13
Paris India Society, 48, 50
Patel, Bhikaiji, 45
Patel, Jaijibai Sorabji, 46
Patel, Maniben, 93, 96
Patel, Sardar, 85
Patel, Sorabji Framji, 46
Patel, Vithalbhai, 94
Peasants in Uniform, 12, 15
Peshwa Baji Rao II, 19

Philippines, 63
Powerful Will, 38
Prafulla Nalini Brahma, 185
Pranakrushna Padhiari, 85
Prasad, President Rajendra, 67, 81, 85
Prasad, Raja Beni, 15
Pride of All Women, 13
Pritilata Waddedar, 135-39
Professor A.L.I., 4
Profullanandini Brahma, 184
Purushottam Trikamdas, 115

Qadir, Birjis, 14, 16
Queen Mary's College, 171
Queen Velu Nachiyar, 25
Queen Victoria, 12, 14
Quit India Movement, 55, 60-61, 86, 108, 112, 119-21, 125, 129-30, 143-44, 193-94

Rai, Lala Lajpat, 47, 75
Raja Chellamuthu Vijayaragunatha Sethupathy, 2
Rajarajeswari Amman Shrine, 26
Rajguru, 153, 183, 187
Rajya Sabha, 96
Ram, Lala Achint, 119
Rama Devi Women's University, 87
Rama Devi's Bari Ashram, 200
Ramabai, Pandita, 99
Rana, Singh Rewabhai, 48
Ranade, Mahadev Govind, 99
Rani Chennamma, 8, 9, 10
Rani Gaidinliu Tribal Freedom Fighters Museum, 180
Rani Gaidinliu, 175-76, 179-80
Rani Kittur Chennamma, 6, 7
Rani Lakshmi Bai, 18-23, 33-34
Rani of Nagas, 179
Rani Sakandimuthal, 2
Ranjit S. Pandit, 108
Rao, Damodar, 20, 22
Rao, Golikeri Krishna, 77
Rao, Krishna, 100
Rao, P.K.N., 171
Rao, Raja Gangadhar, 20
Roosevelt, Eleanor, 109
Rosemary Dinnage, 38
Rustomji Cama, 46-48

Sabarmati Pact, 70
Sahgal, Lakshmi, 169-74
Saigal, K.L., 104
Salt Satyagraha, 60, 70, 80, 83-84, 94-95, 102, 142, 144
Sangeet Natak Academy, 104
Sangoli Rayanna, 9
Sangrami Sibira, 201
Sanjiv Rao, 78
Santi Ghose, 181-184, 186-88, 190
Santi-Suniti, 187
Sarala Devi, 156
Sardar Griha, 78
Sardar Vallabhbhai Patel Memorial Trust, 96
Sarojini Devi Eye Hospital, 71
Sarojini Naidu College for Women, 71
Sarojini Naidu School of Arts and Communication, 71
Sastri, Srinivasa, 99
Satyagraha, 54, 90, 95
Satyavati Devi, 118-121
Savarkar, Veer, 47
Savarkar, Vinayak Damodar, 50

Secret Congress Radio, 114, 115
Sen, Kiranbala, 85
Senapati Bapat, 47
Sepoy Mutiny, 21, 29
Seva Dal, 100
Shah, Nawab Wajid Ali, 13
Shankar Parvati, 104
Shivalingappa, 7, 8, 10
Silambam, 2
Singh, Baba Banta, 63
Singh, Bhagat, 149, 153, 183, 187
Singh, Mann, 63
Singh, Raja Desibaksh, 15
Singh, Raja Man, 15
Singh, S Kesar
Gadhar di dhee-Gulab Kaur (Gadar's Daughter-Gulab Kaur), 65
Singh, Sadoba, 33
Sir Cowasji Jehangir, 50
Sir J. J. Hospital, 116
Sir Tej Bahadur Sapru, 99
Sir Thomas Munro, 9
Sivagangai, 25-27
Sivangangai, 4
South African Indians', 55
Soviet Union, 109
Srivastava, Tara Rani, 128-130
St. John Thackeray, 8, 9
Stuttgart, Germany, 45
Suffragette Movement, 49
Sukhdev, 153, 183, 187
Sunita Devi, 74
Swaraj Ashram, 85

Tamilmatic, 4
Tamrapatra Freedom Fighter Award, 180
Tantia Tope, 20-22
Test of Destiny, 126
Thakkar, Babubhai, 115
Thapar, Sukhdev, 150
The Gram Sevak Press, 87
The Queen Mary High School, 94
Theosophical Society's Annual Convention, 41-42
Tilak Kanya Shala, 78
Tilak Swaraj Fund, 74

Uda Devi, 28-31
Udaiyaal, 3, 4
Umabai Kundapur, 76-81
Uncle Sam's Empire, 103
UNESCO, 105
UNIMA, 105
United Kingdom, 109
United Nations, 167
United Nations Commission on Human Rights, 110
United Nations General Assembly, 109
United States, 64, 133
University of Delhi's Kamala Nehru College, 91
Urmila Devi, 74
Usha, 113-14
Utkal Gaurab Madhusudan Das, 84
Utkal Khadi Mandal, 86

Valeri, 2
Vallabh Vidya Nagar, 96
Vande-Mataram, 48, 58, 60, 115, 126, 195
Varanasi, 19
Varma, B.L., *Jhansi ki Rani*, 36
Varma, Shyamji Krishna, 46-47

Veeramangai, 2, 25
Veerthalapathy, 25
Vellachi, 4
Velu Nachiyar, 1-4, 25-26
Vijaya Lakshmi Pandit, 106-07
Vinoba Bhave's Bhoodan & Gramdan, 86
Virupachi, 3
Vivekananda Seva Award, 180
Vohra, Bhagwati Charan, 150, 153
Vrajkunwerba, 53

Wajid Ali Shah, 14
War-Torn China, 103
West Bengal Legislative Assembly, 159, 190
White Slavery, London, 40
William Hutt Curzon Wyllie, 49
William Wood, 39
Women's India Association, 69
World Craft Council, 104-05
World War I, 50, 162
World War II, 102, 108

Zeliangrong Tribe, 177

*